WORLD BANK TECHNICAL PAPER NO. 409

Commercial Management and Financing of Roads

Ian G. Heggie
Piers Vickers

The World Bank
Washington, D.C.

Technical Papers are published to communicate the results of the Bank's work to the development community with the least possible delay. The typescript of this paper therefore has not been prepared in accordance with the procedures appropriate to formal printed texts, and the World Bank accepts no responsibility for errors. Some sources cited in this paper may be informal documents that are not readily available.

The findings, interpretations, and conclusions expressed in this paper are entirely those of the author(s) and should not be attributed in any manner to the World Bank, to its affiliated organizations, or to members of its Board of Executive Directors or the countries they represent. The World Bank does not guarantee the accuracy of the data included in this publication and accepts no responsibility for any consequence of their use. The boundaries, colors, denominations, and other information shown on any map in this volume do not imply on the part of the World Bank Group any judgment on the legal status of any territory or the endorsement or acceptance of such boundaries.

The material in this publication is copyrighted. Requests for permission to reproduce portions of it should be sent to the Office of the Publisher at the address shown in the copyright notice above. The World Bank encourages dissemination of its work and will normally give permission promptly and, when the reproduction is for noncommercial purposes, without asking a fee. Permission to copy portions for classroom use is granted through the Copyright Clearance Center, Inc., Suite 910, 222 Rosewood Drive, Danvers, Massachusetts 01923, U.S.A.

Cover photos clockwise from top left: "Road construction in Ghana," Ministry of Roads & Highways; "Labor-based road rehabilitation in Lesotho," Labor Construction Unity, Ministry of Works; "Periodic maintenance in Colombia," Ian G. Heggie; and "Microenterprise road maintenance unit in Colombia," Ian G. Heggie.

ISSN: 0253-7494

Ian G. Heggie is a roads adviser and Piers Vickers is a rural transport specialist in the transport division of the World Bank's Transportation, Water, and Urban Development Department.

Library of Congress Cataloging-in-Publication Data

Heggie, Ian Graeme.
    Commercial management and financing of roads / Ian G. Heggie,
Piers Vickers.
        p.   cm. — (World Bank technical paper ; ISSN 0253-7494 ; no.
409)
    Includes bibliographical references (p.).
    ISBN 0-8213-4237-1
    1. Roads—Finance.   2. Roads—Maintenance and repair—Economic
aspects.   I. Vickers, Piers, 1967– . II. Title.   III. Series.
HE336.E3H43   1998
388.1'14—dc21   98-19383
                                                        CIP

Contents

Foreword

The road sector is big business. If main road agencies were publicly listed companies, they would rank among the Fortune Global 500. The Japan Highway Public Corporation manages assets roughly equal in value to those of General Motors, the U.K. Highways Agency is in the same league as IBM and AT&T, while a relatively small road agency like the Roads Department in South Africa is in the same league as Northwest Airlines. Given the size and importance of the road business, it is extraordinary that these agencies are still managed through a government department and financed through general budget allocations—in the same way that governments manage the health and education sectors. They keep their accounts on a cash basis, have no balance sheet, and are subjected to little market discipline. And yet what is often a country's largest business is perfectly capable of standing on its own feet.

Most government departments do not have a commercial orientation, and general budget financing is a failure for commercial undertakings. Government budgets were not designed to finance a major business. Roads are big business and should be managed like a business. They should be brought into the marketplace and put on a fee-for-service basis. In other words, the road sector should be commercialized. This involves creating an arm's-length agency to manage at least the main road network on a commercial basis, introducing an explicit road tariff, making sure that road users pay for extra spending on roads, depositing the proceeds from the road tariff into a road fund, appointing a representative public-private board to oversee management of the road fund, establishing a small secretariat to manage the day-to-day affairs of the road fund, and ensuring that all works financed from the road fund are subject to rigorous technical and financial auditing.

This study is an international edition of a report on managing and financing roads first published as a World Bank technical paper in the Africa Technical Series. The paper has been expanded to include examples of sound management and financing practices from all parts of the world—drawing on examples from industrial, developing, and transition economies—and to include more details on institutional management structures and road funds. Support for this international edition was provided by the Swiss Agency for Development and Cooperation and the New Zealand Consultant Trust.

Jean-François Rischard
Vice President
Finance and Private Sector Development

Abstract

In developing and transition economies 60 to 80 percent of all passenger and freight transport moves by road, and roads provide the main form of access to most rural communities. Yet most of the 11 million kilometers of roads in these countries are poorly managed and badly maintained. Harral and Faiz (1988) estimated that in the 85 countries that had received World Bank assistance for roads, allocations for maintenance had been so low that a quarter of the main paved roads outside urban areas and a third of the main unimproved roads had to be reconstructed. Over the past two decades the situation has generally worsened, with allocations for maintenance often falling below 50 percent of requirements.

Two major initiatives were launched to better understand the underlying causes of such poor road maintenance policies and to explore ways of establishing a secure and stable flow of funds: the Africa Road Maintenance Initiative and the PROVIAL program in Latin America. More modest initiatives are currently under way in Asia and the Middle East. These programs have clarified why roads are poorly managed and underfinanced. Indeed, we can now draw working conclusions about the most effective ways to promote sound policies for managing and financing road networks.

The emerging central concept is *commercialization*: bring roads into the marketplace, put them on a fee-for-service basis, and manage them like a business. This contrasts with the usual procedure of managing roads through a government department and financing them through general budget allocations—in the same way that the health and education sectors are managed and financed. But roads do not need to be managed like a social service. Instead, they can be commercialized by introducing an explicit road tariff for users, making sure the road agency does not siphon funds from other sectors; managing the proceeds from the road tariff through a representative road board; and handling day-to-day management through a small secretariat subject to explicit legal regulations and to technical and financial audits. A number of countries in Africa, Asia, Eastern Europe, Latin America, and the Middle East are implementing such reforms designed to promote commercialization.

Abbreviations

General

ADT	Average daily traffic
AGETIP	Agence d'execution des travauz d'interet public contre le sous-emploi
ESCAP	United Nations Economic and Social Commission for Asia and Pacific
HDM III	Highway Design and Maintenance Model, Version Three
IMF	International Monetary Fund
NGO	Nongovernmental organization
OECD	Organisation for Economic Co-operation and Development
RMI	Road maintenance initiative
SDC	Swiss Agency for Development and Cooperation
VOCs	Vehicle operating costs

Units of measurement

$	U.S. dollars
ECU	European Currency Unit
ESAL	Equivalent standard axles
ESAL-km	Equivalent standard axles times distance traveled
GDP	Gross domestic product
GNP	Gross national product
GVW	Gross vehicle weight
GVW-km	Gross vehicle weight times distance traveled
IRI	International Roughness Index
km	Kilometers
m/km	Meters per kilometer
NPV	Net present value
vpd	Vehicles per day
veh-km	Vehicles times distance traveled

Acknowledgments

This policy study was prepared by Ian G. Heggie, Roads Adviser, and Piers Vickers, Rural Transport Specialist on secondment from the U.K. Department for International Development, under the direction of Jean-François Rischard, Vice President, Finance and Private Sector Development. The study is based on work carried out under regional capacity-building initiatives in Africa, Asia, the Middle East, and Latin America, which have been working to make management and financing of roads sustainable in the long term. The earlier work in Africa was summarized in Heggie (1995b). This international edition of the policy study represents a much expanded version of the earlier technical paper. Work on the international edition was supported by the Swiss Agency for Development and Cooperation, the New Zealand Consultant Trust Fund, and the World Bank.

Substantive inputs to the study were made by Robert Butler (consultant), The Babtie Group (consultants), Heather Chalcraft (FedHaul, Zambia), Peter Collis (U.K. Highways Agency), Csaba Csapodi (Directorate for Road Management, Hungary), Robin Dunlop (Transit New Zealand), Miranda Douglas (VicRoads, Australia), Albert du Plooy (South African Roads Department), In-Hee Nam (Bureau of Public Roads, the Republic of Korea), Martin Fletcher (Transit New Zealand), Matti Hermunen and Jukka Isotalo (Finnish Road Administration), Vilnis Millers (Latvian Road Fund), Raja Nowsherwan (National Highway Authority, Pakistan), Alfonso Olavarrieta (Dirección de Vialidad, Chile), Shunsuke Otsuka (Ministry of Construction, Japan), David Rendall (Transit New Zealand), Richard Robinson (consultant), Dezso Rosa (Directorate for Road Management, Hungary), Adnan Safadi (Ministry of Public Works and Housing, Jordan), Ole Sylte (consultant), Valentin Silyanov (Moscow State Automobile and Roads Technical University, the Russian Federation), Andras Timar (European Bank for Reconstruction and Development), and Gunter Zietlow (International Road Federation/Deutsche Gesellschaft für Technische Zusammenarbeit).

Members of the World Bank's Roads and Highways group also provided helpful support and advice, including Jaffar Bentchikou (AFTT2), Sven-Ake Blomberg (ECSIN), Anders Bonde (ECSIN), Rodrigo Archando-Callao (TWUTD), Christopher Hoban (SASIN), Jeremy Lane (ECSIN), Peter Parker (ECSIN), William Paterson (EASTR), Navaid Qureshi (Resident Mission, Pakistan), Pedro Taborga (ECSIN), Antti Talvitie (OEDST), and Jacques Yenny (ECSIN). The text was edited and formatted by Communications Development, Inc.

The study was reviewed by Robin Dunlop (Transit New Zealand), Kenneth Gwilliam (TWUTD), Jukka Isotalo (Finnish National Road Administration), Henry Kerali (University of Birmingham), Martin Snaith (University of Birmingham), and Mel Quinn (U.K. Highways Agency). The authors thank the reviewers for their comments, which have been incorporated into the text.

Overview

Road transport grew rapidly after World War II, when countries expanded their road networks considerably and built new roads to open up land for development. By the end of the 1980s there were about 11 million kilometers (km) of roads in developing and transition economies. These roads now carry 60 to 80 percent of all passenger and freight transport. They also provide the only form of access to most rural communities. In terms of assets, employment, and turnover, these roads are truly big business. For some developing and transition countries roads are their largest assets, with replacement costs amounting to well over $500 billion.

In spite of their importance, most roads in these countries are managed and financed by bureaucratic road departments in the same way that social services are managed and financed. Traffic congestion is pandemic, and there is a huge backlog of deferred maintenance. In the 85 countries that had received World Bank road assistance during the 1980s, maintenance had been so low that nearly 15 percent of the capital invested in main roads—roughly $43 billion—had been eroded. During the past 20 years these countries spent far too little on capital investment and routine and periodic maintenance. They have been consuming their assets. Restoring only the roads for which it is economically justified to do so and preventing further deterioration will now require annual expenditures of at least $5 billion over the next 10 years. Another $5 billion may be needed to expand and modernize congested road networks and to improve road safety.

The costs of poor road management and inadequate road financing are borne primarily by road users. In rural areas, where roads often become impassable in bad weather, agricultural output suffers. When a road is allowed to deteriorate to poor condition, each dollar deferred on road maintenance increases vehicle operating costs (VOCs) by about $2 to $3. In Africa the extra costs due to insufficient maintenance amount to about $1.2 billion per year, while in Latin America and the Caribbean the cost is $1.7 billion per year. In India VOCs could be reduced by an estimated $4 billion per year through better road maintenance. Moreover, about 75 percent of these costs in developing and transition economies are paid for with scarce foreign exchange. Not surprisingly, road user organizations, particularly those in countries like Jordan, Pakistan, the Philippines, South Africa, Surinam, and Zambia, are willing to pay for roads provided the money is in fact spent on roads and the work is done efficiently.

The Africa Road Maintenance Initiative (RMI), the PROVIAL program in Latin America, and similar country initiatives in Asia and the Middle East have shown that roads are poorly managed and underfinanced because of weak institutional frameworks. Road construction and finance are not market-driven, and there is no clear price for roads, as road expenditures are usually financed from general tax revenues. Roads are procured through appropriations and compete against other claims. Other weaknesses also prevail in the road sector: poor terms and conditions of employment, lack of clearly defined responsibilities, ineffective and weak management structures, and little managerial accountability. A compelling remedy is real or surrogate market discipline, in the form of competition, that motivates road agency managers to cut waste, improve operational performance, and allocate resources efficiently.

The strategic mechanism for promoting competition is commercialization: bring roads into the marketplace, put them on a fee-for-service basis, and manage them like a business. This is not the same as earmarking gen-

1

eral budget revenues as a means of capturing more of the government's overall budget for the road sector. Earmarking has never worked, as is shown in this report. Commercialization is different and requires complementary reforms in four other important areas. These four basic building blocks focus on: clarifying *responsibility* by assigning roles difinitively; creating *ownership* by involving road users in the management of roads to encourage better management, to win public support for road funding, and to constrain spending to what is affordable; stabilizing road financing by securing an adequate and *stable flow of funds*; and *strengthening management* of roads by introducing sound business practices and improving managerial accountability.

Assigning Responsibility

The aim of the first building block is to create a consistent organizational structure with clearly assigned responsibilities for managing different parts of the road network, including road traffic. This requires allocating responsibility among different departments and levels of government, and allocating responsibility between communities and road agencies of the central and local government. To work, these arrangements need an accurate road inventory, a functional classification of roads, designation of appropriate road agencies, formal assignment of responsibility among agencies, and a clear definition of the relationship between the road agency and the parent ministry. Responsibilities to be assigned include operations, maintenance, improvements, road network development, traffic management, accident and claims resolution, and assessment of environmental impacts.

The main trunk road network is usually managed through a central government department, typically the ministry of works. Main roads are costly to build and maintain. With growing traffic, some of these are now operated as toll roads. Many countries make the main road agency responsible primarily for high-volume (national) roads and expressways and are attempting to use the private sector to manage state toll roads or to build and operate them under concession agreements. Rethinking the role of the main road agency has led to the establishment of a growing number of semi-autonomous road agencies that manage the trunk road network on a commercial basis.

Regional road networks are often extensive, and they can thus be managed like the main road network. But this is rarely true of rural road networks—local government entities are often small and lack both technical capacity and resources. The difference is essentially a matter of scale. Countries have attempted to deal with this problem in one of four ways:
• Shifting the legal responsibility for rural roads to a central government department or to a specialized rural roads agency. This tends to improve road conditions initially, but at the expense of further weakening local governments, restricting local input into the planning process, threatening long-term sustainability, and damaging efforts to decentralize responsibility.
• Establishing a project implementation agency to plan and implement road projects on behalf of local governments. This solution works, although current arrangements are not subject to competitive bidding and are overly reliant on donor funding.
• Bringing several local government agencies together to procure goods and services collectively. These arrangements, known as joint-services committees, offer many advantages, though the agreements tend to be complex.
• Contracting out the planning and management functions to consultants who may work for several local government road agencies. The local government agency remains responsible for the roads, but planning and management is delegated to a technically qualified third party.

How urban roads are managed depends essentially on the size of the urban area. In large metropolitan areas roads may be managed by a citywide authority that has total responsibility for all roads and highways; the city's different political jurisdictions, which have total responsibility for the roads and highways within their own jurisdictions; a strategic authority—an elected authority or a regional or state body—that takes responsibility for certain strategic roads or functions but leaves all other roads and highways under local jurisdiction; or a strategic authority made up of representatives from each local jurisdiction. Works may be implemented by each body using its own directly employed work force and/or contractors, or it may be implemented by the strategic authority, which then becomes responsible for delivering the services to each jurisdiction.

Large independent towns and cities may manage their road networks along the same lines as in large metropolitan areas. Typically the urban area has total responsibility for its own roads, receiving varying inputs from local, regional, or national governments. The urban area is usually free to deliver services however it sees fit, although it may be subject to guidelines laid down by the region or national government. The tendency today is to use contractors for most of the work. In small urban areas where the urban government often lacks the scale necessary to manage its own road network, responsibility for managing roads may rest with a regional or national agency. Alternatively, the urban authority may be responsible for managing nonstrategic roads or carrying out a limited range of functions (for example, street cleaning, pothole repairs), while a regional or national agency is responsible for strategic roads or larger-scale works. Most core functions will be done under contract.

Managerial responsibility at the lowest level of the road network is poorly defined. Often agencies have few technical skills and scarce funding. Some countries treat the problem by creating a special class of roads funded on a cost-share basis. Local inhabitants are given incentives to assume ownership of the roads and organize and manage road cooperatives, which are at least partially financed by the central and local government. The community contribution may be in the form of cash or volunteer labor, with oversight and technical advice provided by the rural road agency.

Regulatory responsibilities (design standards, signs and signals, parking, congestion, routing of heavy vehicles) for all types of roads are usually assigned to the main road agency, though these may be delegated to other road agencies or competent bodies, especially responsibilities that are likely to have a significant urban impact. Responsibility for enforcing axle-weight regulations is generally assigned to the main road agency, ideally with the cooperation of the road transport industry. The main road agency may also carry out vehicle safety and vehicle emission inspections and assess environmental impacts.

Ensuring Ownership

This second building block requires the active participation of road users to help win public support for secure and stable road funding. But support for more road funding through a user-pay or fee-for-service arrangement requires that steps be taken to ensure that road agencies do not operate as public monopolies and that no more is spent on roads than the country can afford. It is thus critical to involve road users in road management—a precondition for getting them to pay for roads willingly. What is needed is a partnership between road users and government to strengthen road management and raise appropriate finances. At the national and regional level road users can be effective by serving on popularly constituted and representative road management boards, as in Finland, Ghana, Malawi, South Africa, Sweden, and Zambia. Most of these boards use outreach programs to keep their constituents and the public informed about the status of the road sector and its management.

Maintaining Steady Financing

The third building block seeks adequate and stable funding. Most governments cannot increase budget allocations given present fiscal conditions. Improved means for mobilizing revenue are essential. Several countries are addressing this issue by separating road financing from the government's consolidated budget. They have introduced an explicit road tariff consisting primarily of vehicle license fees and a fuel levy. The revenues are collected independently of the government's sales and excise taxes and are deposited directly into a road fund. In some cases the road tariff is collected under legislation that defines it as a tariff rather than as part of the government's tax revenues (that is, such road funds are road public utilities). The tariff is generally set to cover the full cost of operating and maintaining main roads and part of the cost of operating and maintaining urban and rural roads. Some countries finance only some rehabilitation and minor new works through the road tariff, while others finance all road spending.

The road fund should be managed by a representative public-private board. At least half the members should come from outside government—chambers of commerce, the road transport industry, farmers organizations, and professional institutions that have a strong vested interest in efficiency and honesty. The day-to-day affairs of the road fund are then managed by a small secretariat headed by a board-appointed chief executive officer (CEO). Legal regulations govern management of the road fund, requiring regular technical and financial audits. This system is, in effect, commercial management of the road fund. It currently prevails in countries like Latvia, Lesotho, Malawi, New Zealand, South Africa, and Zambia

Promoting Commercial Management

The final building block calls for the creation of a businesslike road agency. Road users involved in the management of roads generally press for sound business practices to ensure value for money. They expect a clear, unambiguous corporate mission and a strategy to separate planning and management of road works from implementation. This may involve contracting out implementation to the private sector, learning effective ways of contracting out, recruiting and paying capable staff, and building sound management structures and appropriate management information systems. These reforms improve market discipline, give managers the freedom to operate commercially, and strengthen man-

agerial accountability. They also encourage objectivity in setting priorities, adopting quality assurance programs, comparing in-house work to that done by contractors, and evaluating appropriate technology for road works. Finally, auditing procedures also must be improved to ensure that the public gets value for money from road spending.

***

These four building blocks represent the core reforms. They are interdependent and, ideally, should be implemented in a complementary fashion. All are necessary. The system for managing and financing roads cannot be reformed until responsibilities are clearly established. The financing problem cannot be solved without the strong support of road users. The support of road users cannot be secured without ensuring that resources are used efficiently. And resource use cannot be improved without controlling monopoly power, constraining road spending, and increasing managerial accountability.

There is, however, scope for flexibility. The reforms can be introduced in different ways, and the content of each building block can differ depending on country circumstances. Reforms can move sequentially or in parallel, and both sequencing and pace can vary. But in the end all four building blocks must be in place to ensure that the reform agenda is sustainable.

Part I.
Background

1. Introduction

This report is an international edition of an earlier World Bank technical paper on managing and financing roads (Heggie 1995b). The technical paper was well received, and the World Bank was asked to produce a revised and expanded version suitable for use in all of the World Bank's regions: Africa, East Asia and the Pacific, South Asia, Europe and Central Asia, Middle East and North Africa, and Latin America and the Caribbean. The technical paper was therefore expanded to include examples of sound management and financing practices from all parts of the world—industrialized, developing, and transition economies—and more details on institutional management structures and road funds.

The report also follows up on the World Bank's policy study, *Road Deterioration in Developing Countries* (Harral and Faiz 1988). According to that study in the 85 countries that had received World Bank assistance for roads, allocations for road maintenance had been so low that nearly 15 percent of the capital invested in main roads—roughly $43 billion, or about 2 percent of these countries' GNP—had eroded because of poor maintenance.[1] As a result a quarter of the main paved road network, together with a third of the main unimproved network, had to be reconstructed. Reconstruction—costing $40 to $45 billion worldwide—could have been avoided by spending only $12 billion on preventive maintenance. The study also argued that if countries did not improve road management, the eventual costs of restoration would increase two to three times, and the vehicle operating costs (VOCs) by even more.

There are several reasons for this calamity. Road authorities were not directly affected by road deterioration and came under no immediate pressure to prevent it. Road users, on the other hand, were slow to see the link between poor road conditions and higher VOCs and, even when they did, were rarely organized to act. The cause of the problem was a lack of public accountability—which could not be solved by additional financial resources alone. The institutional base of the road sector had to be reformed, including organization, staffing, and performance.

Harral and Faiz (1988) offered few specific solutions but did give some direction. It pointed out that road agencies were usually public monopolies and had too many responsibilities, including planning, controlling, and executing construction and maintenance programs. Furthermore, they devoted too much staff time, funds, and facilities to executing road works. In most countries planning, controlling, and executing should be separated and the execution of road works should be transferred to the private sector or to a specialized government construction agency, so as to clarify responsibilities, improve incentives, and strengthen accountability. Road agencies also needed better management information systems to better plan their investment and maintenance programs. Finally, the study argued that internal accountability had to be improved, perhaps by mobilizing the media and nongovernmental organizations (NGOs) to help politicians and the public become aware of the high costs of insufficient maintenance.

Harral and Faiz (1988) was an important milestone in the debate on road maintenance policies. It gave impetus to a number of initiatives designed to understand the underlying causes of poor road management. It also encouraged road agencies to address these institutional issues through a clearly articulated reform program. The Road Maintenance Initiative (RMI), a major component of the Sub-Saharan Africa Transport Policy Program, was one of these initiatives, as was the

PROVIAL ("for roads") program in Latin America. These programs did a great deal to improve the understanding of why roads were poorly managed and underfinanced. Indeed, we can now draw tentative conclusions about the most effective way to promote sound road management policies and the broad outline of the policies themselves. This report summarizes the lessons learned from these programs, combines them with lessons learned from industrial countries, and then develops an overall agenda for reforming road management and financing.

This report is written for a nontechnical audience and is directed at country policymakers, World Bank management and staff, officials in other development agencies, and senior officials in developing and transition economies—all those interested in improving management and financing of roads and making them sustainable in the long term.

Note

1. The specific figures were: Africa, $5.0 billion; East Asia and the Pacific, $9.3 billion; South Asia, $8.6 billion; Europe, the Middle East, and North Africa, $9.3 billion; and Latin America and the Caribbean countries, $11.0 billion. These figures excluded estimates for most of the former Soviet Union.

2. The State of the Road Sector

This chapter argues that, in spite of the economic and financial importance of roads in developing and transition economies, most are poorly managed and badly maintained. The chapter then examines the economic impact of poor road maintenance policies, reviews past attempts to reform them, and outlines some of the regional initiatives put in place to try to solve them.

The Importance of Roads and Road Transport

Road transport grew rapidly after World War II and is now the dominant form of transport throughout the world. The importance of main road networks is gauged by the proportion of total passenger and freight movement made by road and by the size of the road business.

Most economies now rely heavily on road transport for passenger and freight movement. In Latin America and the Caribbean road transport accounts for more than 80 percent of domestic passenger transport and more than 60 percent of freight movement. In Africa these proportions are even higher. In the countries of the former Soviet Union roads account for 80 percent of freight and 58 percent of passenger transport. In the United States 34 percent of inland surface freight is now transported by road, and 98 percent of inland surface passenger transport is made using private cars.[1] Respective figures for the Republic of Korea are 38 percent and 76 percent, and for Egypt are 80 percent and 46 percent.

In almost all countries the proportion of passenger and freight transport carried by road is increasing, rapidly in some. Traffic in Thailand, for example, has grown 14 percent annually since 1986. Even countries historically dominated by other modes of transport are now witnessing remarkable expansion in demand for road transport. In the Russian Federation, which has relied largely on the railway network, the total freight moved by road is expected to increase from 13 percent to between 22 and 41 percent in the coming years (depending on the rate of economic growth)—and Russia is currently the second largest freight mover in the world. The hard truth is that roads are the main arteries for moving goods and people in the global economy, and they are becoming increasingly dominant. But such growth in demand for road space is stressing road networks not designed to carry such high volumes of traffic, especially heavy vehicles.

The Size of the Road Business

The road business itself is massive in terms of human-made assets, investment, and revenues. In response to rapid traffic growth, countries expanded their road networks considerably, particularly during the 1960s and 1970s. They also built new roads to open up more land for development. Expansion has been especially fast in Asia. The road network in Korea has grown threefold in the past 35 years—to its present length of more than 77,000 km. During 1984–94 the road networks in Indonesia, Korea, Malaysia, and Pakistan grew in length by more than 5 percent per year (ESCAP 1995). Most other countries also witnessed road network expansion. For example, Jordan had only 895 km of roads in 1950 but more than 7,000 km by 1996. Today there are nearly 1.5 million km of roads in Africa, 3 million km in Central and South America, 2.6 million km in Asia (excluding China and India), just under 1 million km in the former Soviet Union, and 500,000 km in the Middle East. Recent estimates have put the asset value of the African road network at more than $150 billion, the network in Latin America and the Caribbean at more than $200 billion, and the Indian highway network alone at $15 billion.

Investment sums are colossal. The World Bank supported more than $5.5 billion of highway and rural road projects in 1994–97. The lion's share has gone to East Asia ($2.2 billion), Latin America and the Caribbean ($1.1 billion), and Eastern Europe ($0.8 billion). The Inter-American Development Bank has been lending approximately $550 million per year, and the Asian Development Bank at least $750 million per year, to support road programs in their respective regions. The industrial world is also not immune to pressures to build roads. Between 1991 and 1995, for example, the European Investment Bank lent ECU 9.6 billion for roads and highways to countries within the European Union, making roads the biggest sector in terms of volume lent. (The above figures are from annual reports of the World Bank, Inter-American Development Bank, Asian Development Bank, and European Investment Bank.) In 1991 Japan invested $29 billion in new roads, and the United States $6 billion (OECD 1994).

The fundamental change in attitude toward the role of the state and the enormous demand for resources in the road sector have forced governments to turn to the private sector to finance a small but significant portion of total road investment. At least 30 countries have adopted private sector road concessions, including six in Latin America and the Caribbean, three in Eastern Europe, and eight in Southeast Asia. The private sector invested about $18.8 billion in 40 road projects in developing and transition economies between 1982 and 1994—more than two and a half times the value of private investment in ports, railways, and airports combined. It is important to remember, however, that private finance focuses on heavily trafficked roads and thus applies only to a small part of the overall road network. For example, in China only 11 percent of the national trunk highway system, itself a small fraction of the total network, is estimated to generate enough toll revenue to attract 100 percent foreign private finance, while a further 36 percent could be funded by less costly domestic loans.

Although experience in many industrial economies suggests that road building can never outpace the growth in demand for road space, some countries are responding to the predicted large growth in traffic by investing in limited-access, frequently tolled expressways. The trend is most apparent in East Asia, Latin America, and Eastern Europe. In partial answer to Korea's explosive increase in traffic (it is predicted that by 2001 there will be 24 million vehicles in Korea, one for every member of the working population) and growing environmental pollution, the Korea Highways Corporation is constructing a grid-shaped expressway network. The plan calls for building 1,800 km of new expressways and upgrading a further 500 km at a cost of about $30 billion. Once the grid is built, no two places in the country will be more than half a day's journey apart. Similar plans include the Trans Java Tollway System in Indonesia, which aims to construct a further 310 km of new expressways by 2000, financed largely through private investment, and an intercity motorway program in Thailand, which has been constructing major arterial routes since 1988, its goal being more than 4,000 km.

Road Revenues and Financing

The importance of roads is further reflected by the fact that spending on roads can absorb as much as 5 to 10 percent of a government's recurrent expenses and 10 to 20 percent of its development budget. Of course, road transport can also make up one of the largest contributions to fiscal revenues. For example, road-user taxes and charges in the United States amounted to $78 billion in 1994 (6.2 percent of federal government revenue) and in the United Kingdom $33 billion in 1995–96 (of which only $10 billion was spent on roads). Net fiscal flows from the road sector tend to be positively correlated with economic development.

Many of the poorest countries in the world continue to subsidize road transport through the general budget.[2] Furthermore, in many countries a significant proportion of the central government's disbursed and outstanding debt is loans made for roads. The road sector also absorbs a great deal of grant finance, mainly for procuring construction and maintenance equipment. Even a relatively small national road agency often owns $25 to $50 million of plant and equipment.

Thus in terms of assets and turnover, particularly when maintenance is fully funded, the world's roads are truly big business—generally bigger than railways or national airlines (table 2.1). Road maintenance and construction is also significant in terms of employment, although privatization and devolution have led to sharp reductions in the number of workers employed in the road sector.

Roads in Rural Areas

Although it is now accepted that roads are not enough to overcome the transport burden of the poor, rural roads can, together with the necessary means of transport, provide significant economic and social benefits in rural areas. While the case for investing in new rural roads is controversial and complex and must be judged country by country, the case for maintaining reasonably trafficked rural roads is clear. A substantial body of evidence from many countries demonstrates the catalytic role of rural roads in agricultural development. Creightney (1993b) summarized the evidence, concluding that transport infrastructure spending often improves productive activity in the agricultural sector.

Roads can also confer social benefits by providing rural communities with access to markets, services, employment, and information. World Bank (1996b) found that improving roads had a significant effect not just in reducing VOCs, which then translated into less expensive public transport, but also in expanding public services provided in rural areas. As a result educational enrollment grew substantially, and residents were able to visit health professionals more regularly.

The Impact of Poor Road Maintenance

Roads in many parts of the world are poorly managed and badly maintained, usually by bureaucratic government road departments. The poor state of the road network is reflected in the large backlog of deferred maintenance. In Africa alone it would cost nearly $43 billion to fully restore all roads that are classified as in poor condition (that is, requiring immediate rehabilitation or reconstruction). In Latin America and the Caribbean the most recent rough estimate (1992) held that it would cost about $2.5 billion per year for a decade to remove the backlog and prevent further accumulation of deferred maintenance. The estimate of the accumulated backlog of maintenance in Kazakhstan is $1.8 billion, while that in Russia for the federal highway network alone is $4.5 billion per year

Table 2.1 Assets, employment, and turnover for roads, railways, and airlines in selected countries, 1995–97
(millions of dollars)

	Chile	Ghana	Hungary	Indonesia	Jordan	Korea, Rep. of	Russian Federation	South Africa	Uruguay
Main and provincial road agencies									
Total assets[a]	4,144	1,665	4,238	11,148	820	4,870	41,963	21,017	1,080
Staff (number)	4,265	3,935	5,598	20,100	8,000	1,450	2,000[b]	330[c]	3,378
Turnover[d]	526	189	348	1,267	74	1,498[e]	1,506	874	94
National railways									
Total assets[f]	537	—	3,968	545	65	8,062	—	7,333.9	118
Staff (number)	5,000	6,081	72,429	35,671	1,223	37,068	1,694,400	64,682	2,115
Turnover	63	7	413	157	11	1,493[e]	5,064	1,900	28
National airlines									
Total assets[f]	79	negligible	124.1	31	970	6,296	880	861	negligible
Staff (number)	2,255	1,164	3,465	14,503	4,796	16,518	13,362	1,056	817
Turnover	173	negligible	297	1,466	392	3,342	1,051	1,166	negligible

— Not available.

a. Based on replacement costs of existing road network with allowance made for road condition. Calculations are available in annex 1.

b. Federal road agency staff only. There are 87 regional road agencies, some of which have several thousand employees.

c. National road agency staff only. Provincial numbers are unreported.

d. Total expenditures on the main and secondary road networks. Country variations are mainly due to variation in the length of paved roads.

e. Not including $7.5 billion in new construction.

f. Based on the replacement costs of total fixed assets or the replacement costs estimated from historic costs.

Source: Country road agency survey; annex 1; World Bank sector and project reports; World Bank task managers; International Air Transport Association, World Air Transport Statistics; World Bank Railways Database.

over an unspecified period. Even in industrial economies such backlogs are common and becoming more so. For example, in 1996 a survey conducted by the U.K. Institution of Civil Engineers found that in Great Britain there was a $5.61 billion maintenance backlog on local government roads (96 percent of the total public network). Because countries have consistently spent far too little on routine and periodic maintenance in the past 20 years, much of the large amount of money already invested in roads has been eroded.

Who Pays for Poor Maintenance?

The economic costs of poor road maintenance are borne primarily by road users. When a road is allowed to deteriorate from good to poor condition, each dollar saved on road maintenance increases VOCs by between $2 and $3.[3] Far from saving money, cutting back on road maintenance increases the cost of road transport and raises the net cost to the economy as a whole. Furthermore, when traffic levels rise, as they have been in most countries, the proportion of total road transport costs attributable to vehicle operation will also increase sharply, while those attributable to road expenditures will decline.

It is estimated that the extra costs of insufficient maintenance in Africa amounts to about $1.2 billion per year, or 0.85 percent of regional GDP. In Latin American and the Caribbean equivalent figures were estimated at $1.7 billion per year in 1992, amounting to 1.4 percent of individual countries' GDP. The Ministry of Surface Transport in India has estimated that $4 billion of the roughly $39 billion in annual VOCs could be saved through proper road maintenance—more than twice total annual expenditures on capital and maintenance works on national and state roads (Indian Ministry of Surface Transport 1996). About 75 percent of these additional VOCs in developing and transition economies must be paid with scarce foreign exchange. It is no surprise that road maintenance and rehabilitation projects produce economic rates of return in excess of 35 percent.[4]

Four maintenance strategies evaluated in 33 countries have proven to be highly cost-effective, with annualized benefit-cost ratios varying from 1.4 to 44.8 (box 2.1). In other words, on an annualized basis each dollar spent on patching and overlays saves at least $1.4 in operating costs and can save as much as $44 depending on traffic volume.

Though based on the roughness of road pavement, the analysis does not fully reflect pothole damage. Most vehicles, particularly loaded freight vehicles, are not designed to deal with the sharp, repeated shocks caused by potholes. Trucking companies are well aware of the extra costs that poor roads impose on road transport operations. According to a recent Russian study, trucks operating on unmaintained rural roads during harvest time suffer a staggering 30 percent reduction in vehicle life, with a resulting sharp increase in depreciation and hence VOCs.[5] An unpublished study conducted in 1992 by the Federation of Zambian Road Hauliers estimated that the additional costs associated with potholes amounted to more than $14,000 per truck per year in spare parts alone—an increase in VOCs of 17 percent. Furthermore, this figure did not include extra fuel, accidents, down-time for repair, and damage to freight inside the vehicle. It is no wonder road transport associations around the world keep pressing for better road maintenance and express a willingness to pay for it—provided that the money is spent on roads and that the work is done efficiently. This common-sense view, well-supported empirically, is at the heart of a functional maintenance policy.

What are the Long-Term Costs?

Poor road maintenance also raises the long-term costs of maintaining the road network. Maintaining a paved road for 15 years costs about $60,000 per km. If the road is allowed to deteriorate over the 15-year period, it will cost about $200,000 per km to rehabilitate it. In other words, rehabilitating paved roads every 10 to 20 years is more than three times as expensive, in cash terms, as maintaining them on a regular basis, and 35 percent more expensive in terms of net present value discounted at 12 percent per year.

For example, a recent unpublished transport sector review for the Republic of Kazakhstan analyzed how the absence of periodic maintenance, due to insufficient funds, affected the national road network. The analysis demonstrated that if periodic maintenance or strengthening was deferred for four years on 7,000 km of roads, the government would save $180 million per year in maintenance costs, but would then have to

Box 2.1 The impact of road maintenance on vehicle operating costs

This example analyzes the impact of road maintenance on VOCs using data from 33 countries. It compares a limited number of potential road maintenance strategies against a base case consisting of routine maintenance only (that is, off-carriageway work). The four maintenance strategies evaluated are:

• Patching, plus 5 cm overlays when surface roughness reaches 6.0 international roughness index (IRI) meters per kilometer (m/km).
• Patching, plus 5 cm overlays when surface roughness reaches 5.0 IRI (m/km).
• Patching, plus 5 cm overlays when surface roughness reaches 4.0 IRI (m/km).
• Patching, plus 5 cm overlays when surface roughness reaches 3.0 IRI (m/km).

The evaluation looked at these strategies over a 50-year period during which traffic was assumed to grow at 3 percent per year. The benefits and costs of each option were calculated using a 12 percent discount rate.

The results are summarized below for roads in fair condition for average daily (two-way) traffic (ADT) volumes of 300, 1,000, 3,000, and 10,000 vehicles per day (vpd). Thirty percent of the traffic consists of trucks with medium loading (that is, the loading corresponds to the average loading for the 33 countries included in the analysis). To make the tables understandable to a wider audience, expenditures on maintenance and VOC savings have been expressed as equivalent annual discounted outlays divided by savings, rather than as total net present value. The benefit-cost ratio thus shows the equivalent annual (discounted) payoff from each strategy.

Road maintenance is shown to be highly cost-effective, with equivalent annual benefit-cost ratios that vary from 1.4 when traffic volumes are 300 vpd to 44.8 when traffic volumes are 10,000 vpd. That is, each equivalent annual dollar spent on maintenance saves at least $1.4 per year in VOCs (with 300 vpd) and as much as $44.8 per year (with 10,000 vehicles per day).

	ADT = 300 vpd				Fair condition, ADT = 1,000 vpd			
Strategy:	1	2	3	4	1	2	3	4
Increased maintenance (dollars per year)[a]	2.39	4.83	7.96	10.15	2.72	4.94	8.04	10.13
VOC savings (dollars per year)[b]	3.32	4.74	5.88	6.15	12.48	16.83	20.69	21.59
Benefit-cost ratio[c]	1.39	0.98	0.74	0.61	4.59	3.41	2.57	2.13
Net present value (millions of dollars)	8.69	–0.85	–19.31	–37.26	90.73	110.63	117.62	106.59
Incremental benefit-cost	n.a.	0.58	0.37	0.12	n.a.	2.04	1.24	0.43

	ADT = 3,000 vpd				ADT = 10,000 vpd			
Strategy:	1	2	3	4	1	2	3	4
Increased maintenance (dollars per year)[a]	4.07	5.82	8.42	10.51	3.88	5.68	8.32	10.08
VOC savings (dollars per year)[b]	56.02	67.26	78.05	80.86	173.83	213.54	250.51	258.79
Benefit-cost ratio[c]	13.76	11.55	9.27	7.69	44.84	37.60	30.13	25.69
Net present value (millions of dollars)	483.14	571.40	647.64	654.28	1,580.7	1,933.31	2,252.68	2,313.33
Incremental benefit-cost	n.a.	6.42	5.16	1.34	n.a.	22.06	14.00	4.71

n.a. Not applicable.
a. Equivalent annual VOC savings attributable to increased maintenance spending.
b. Equivalent annual expenditures in addition to routine maintenance.
c. VOC savings divided by spending on increased maintenance.

spend $1.05 billion reconstructing the roads. Thus the net loss would be at least $330 million.

The same pattern holds for gravel roads. Maintaining a gravel road for 10 years costs between $10,000 and $20,000 per km, depending on climate and traffic volume. But leaving it without maintenance for 10 years will cost about $40,000 per km for needed rehabilitation. Rehabilitating gravel roads every 10 years is thus twice as expensive, in cash terms, as regular routine and periodic maintenance, and between 14 and 128 percent more expensive in terms of net present value discounted at 12 percent per year.

In rural areas, where roads often become impassable during bad weather, poor road maintenance profoundly affects the economy. Poor maintenance can result in large, direct economic costs in terms of lost production. Crops and other agricultural produce often spoil for want of a passable road to take perishable products to market. For example, in 1988 the Deputy Prime Minister of Russia suggested that in areas where agriculture was of secondary importance, poor roads resulted in losses of up to 15 percent of agricultural production.

The lack of maintenance can also seriously impair people's lives in social terms, when roads become impassable and communities can no longer access markets and public services, particularly emergency health care. While such social losses have not yet been measured empirically, they are likely to be considerable and affect a large proportion of the world's poorest population.

Past Efforts at Reform

Until the beginning of the 1990s most reform efforts sought to strengthen road management, improve policies governing user charges, and increase allocations for road maintenance. But these reforms lacked a comprehensive vision focused on technical rather than institutional solutions, and were generally implemented in piecemeal fashion.

Some attempts were made to rationalize and decentralize road management, but little effort was made to deal with weaknesses in the organizational structures of road agencies, low pay scales, shortages of qualified staff, lack of staff motivation, and lack of managerial accountability. Instead, most reforms concentrated on reducing work done using in-house staff and equipment, introducing maintenance management systems, and restructuring government plant and equipment pools. These initiatives were accompanied by complementary attempts to simplify government procurement procedures so as to facilitate the use of local contractors, strengthen the local construction industry, introduce maintenance and equipment management systems, and strengthen axle-weight enforcement to reduce the damage overloaded vehicles inflicted on road pavement. The most successful reforms dealt with work done using in-house staff and equipment, sim-

plifying procurement procedures, and strengthening the local construction industry. The remaining initiatives had little lasting impact because of shortages of qualified staff, managerial indifference, and resistance from strong vested interests.

Reforms of user-charging policies encouraged governments to adopt charges based on short-run marginal costs, that is, variable road maintenance costs and road congestion costs.[6] The aim was to encourage best use of the road network and ensure that heavy vehicles paid for the damage they did to the road pavement. These efforts were partly successful. Taxes on heavy vehicles were often increased following studies of road-user charges, but no country was willing to accept strict short-run marginal cost pricing for roads. Governments did not see the point of using such a pricing system on uncongested roads, why road users should be subsidized by other sectors of the economy, and how the proposed arrangements made fiscal sense. With little road congestion, user charges would be set equal to variable road maintenance costs, which would cover only about half the costs of operating and maintaining the road network.

Attempts to improve road financing concentrated on increasing allocations for road maintenance and, in Africa and Latin America, earmarking funds to secure a stable flow. Donor countries often asked governments to set aside part of their general tax revenues (usually specified as a percentage of overall fuel tax revenues), deposit the money into a road fund, and use the proceeds to finance maintenance of the core road network.

But apart from pointing out the economic costs of deferred maintenance and suggesting that funds be reallocated from construction to maintenance, little advice was offered on where the additional revenues might come from and how the road fund should function. The International Monetary Fund (IMF) opposed earmarking on grounds that it undermined unified budget management. Ministries of finance objected to road funds on grounds that they simply did not work (see de Richecour and Heggie 1995). Thus most road funds suffered from systemic problems—deposits were erratic, withdrawals were frequently delayed, governments diverted money to finance other public programs, and expenditures were loosely controlled—and failed to provide an adequate, stable flow of funds.

Regional Programs

Against this background two regional programs have sought to give more focus to road sector reform. The first and most successful is the Road Maintenance Initiative (RMI), renamed the Road Management Initiative in April 1997. The United Nations Economic Commission for Africa and the World Bank launched this program in the late 1980s as one of the five components of the Sub-Saharan African Transport Policy Program. The second regional program is PROVIAL—established in 1992 through an initiative of the World Bank's Economic Development Institute (EDI) to address road maintenance management in Latin America.

Road Maintenance Initiative in Africa

RMI has sought to identify the underlying causes of poor road maintenance policies and to develop an agenda for reforming them within Africa. Relying primarily on subregional seminars, the initial phase of the RMI program raised awareness of the need for sound road maintenance policies and identified why current approaches were ineffective and unsustainable. The second phase then encouraged country initiatives in Kenya, Madagascar, Nigeria, Tanzania, Uganda, Zambia, and Zimbabwe. The country programs focused initially only on main roads and concentrated on promoting reforms in three main areas: planning, programming, and financing; operational efficiency; and institutional and human resource development.

During the initial stages of the program the policy dialogue quickly led to three important insights. First, it had always been assumed that the ministry of finance would play a key role in developing sustainable road maintenance policies. So strong was this belief that some country initiatives sought to interest the ministry in road maintenance by exploring the basic financial issues through public expenditure reviews. But it quickly became apparent that involving the private sector was the secret to success, the ministry of finance did not hold the key. The private sector, after all, used and paid for the roads and clearly had the most to win or lose. Their representative organizations—chambers of commerce, road freight and passenger transport associations, and agricultural organizations—were strong and influential. Their support could often overcome bureaucratic resistance, whether from the ministry of works or the ministry of finance.

Second, many systemic problems associated with poor road maintenance policies—weak programming and budgeting, undue emphasis on work done using in-house staff and equipment, and inefficient plant pools—were merely symptoms of an underlying institutional problem. The real problems were weak or unsuitable institutional arrangements for managing and financing roads and the impact these arrangements had on staff incentives and motivation, as well as on managerial accountability. Until the institutional framework is strengthened, it will be almost impossible to overcome the numerous technical, organizational, and human resource problems that hamper sound road maintenance policies.

Third, attempts to improve road maintenance policies cannot be limited to maintenance alone or to the maintenance of main roads. Poor road maintenance policies are a subset of the wider issues of managing and financing roads. Moreover, the problems tend to be most serious at the regional and district levels, where institutional weaknesses are more acute and finances scarce.

These insights opened the two-way dialogue between the RMI program and the participating countries to a wider debate about the institutional arrangements for managing and financing all types of roads. RMI's message has been disseminated through various media, including regional and subregional policy seminars, country workshops, study tours, annual meetings of all country representatives and participating donors, newsletters, and visits by RMI staff. The result was a program that helped to promote a number of major policy reforms in several African countries.[7]

PROVIAL in Latin America

EDI took the lead in establishing PROVIAL, though it collaborated with a number of bilateral and multilateral agencies including the International Road Federation (IRF), the Permanent International Association of Road Congresses (PIARC), the U.S. Federal Highways Administration, and the German Gesellschaft für Technische Zusammenarbeit (GTZ). The objectives of the PROVIAL program were similar to that of the RMI: creating awareness of the need for proper road maintenance, encouraging adequate and timely funding for road maintenance, promoting the concept of accountability in government, and encouraging the transfer of results from research and development to improve managerial and construction techniques.

PROVIAL has relied heavily on seminars to disseminate its message and to enable a dialogue among Latin American and Caribbean countries and between Latin American and Caribbean countries and the donor community. Between 1992 and 1995 PROVIAL organized 17 seminars, seven of which were regional and the rest held for specific countries. The last such seminar, held in Puerto Rico, concluded that PROVIAL had effectively enabled engineers and technicians from the region to share professional know-how. But it had made insufficient progress in improving road maintenance management. Public debate had to be expanded to involve pertinent public institutions and road users in the reform and decisionmaking process. Country representatives suggested that PROVIAL address other issues, such as consensus building, decentralization of road network management, identification of appropriate financial instruments, options for public-private partnerships, road safety and environmental concerns, and ways to harmonize regional legislation and regulations.

Other Seminars

A number of regional seminars have taken place apart from specific programs such as RMI and PROVIAL. In May 1995 the World Bank, the European Union program for Technical Assistance to the Commonwealth of Independent States (TACIS), and GTZ held a highway policy seminar in Moscow for countries of the former Soviet Union. This seminar provided a forum to exchange experiences, enable the international donors to develop a clear understanding of existing country policies, lay the groundwork for policy reform, and instruct participants on how to prepare loans for international donors. In September 1996 a joint UN Economic and Social Commission for Asia and the Pacific (ESCAP)-World Bank seminar was held in Bangkok, sponsored in part by GTZ and the Swiss

Development Cooperation. This meeting aimed to share international experience with ESCAP member states and to consider whether reforms from other regions might be applicable to Asia and the Pacific, and, if so, how they might be adapted to the Asian environment. By 1997 a number of country workshops had been held in Bangladesh, Pakistan, and the Philippines and several more were planned for India, Laos, and Nepal.

Notes

1. Figures for the former Soviet Union refer to 1993, while those for the United States refer to 1995. See International Road Federation (1997).
2. This relationship was demonstrated in an analysis of government tax revenues that subdivided into general revenue taxes and specific charges for usage of roads in eight countries. See Heggie (1991b).
3. A paved road in good condition, carrying about 500 vehicles per day, requires resealing or light overlays, costing about $23,600 per km, every seven years to keep it in good condition. The net present value of this amount, discounted at 12 percent over 25 years is $17,688 per km. Without maintenance, the road will deteriorate from good to poor condition. This will increase VOCs by about $5,000 per km, which has a net present value, when discounted over 25 years, of $39,200 per km (Thriscutt and Mason 1991, p. 29–30). The benefit-cost ratio of a fully funded road maintenance program is thus between 2 and 3.
4. A recent analysis of the World Bank's Operations Evaluation Department database, covering 341 road projects evaluated between 1961 and 1988, found that the average economic internal rate of return for pure road maintenance projects was 38.6 percent. The analysis was carried out for World Bank (1994).
5. Personal communication from Professor Leo Rothenburg, University of Waterloo.
6. Under this practice the price is made equal to short-run marginal costs (that is, the costs of producing the last unit sold, plus a mark-up to clear the market). The rationale was that subject to certain assumptions about production costs and other matters, such a pricing rule would maximize economic welfare. See, for example, Churchill (1972) and Walters (1968).
7. The RMI was one of the first two winners of the President's Award for Excellence within the World Bank. That award recognizes outstanding achievement in the field of operations.

Part II.

The Basic Issues

3. A Diagnosis

Why have most governments been pursuing ineffective and unsustainable road maintenance policies? There is no simple answer. But despite significant country and regional variations, there are some common themes. The main problems are institutional, affecting incentives. As part of *World Development Report 1994* (World Bank 1994) a survey of 44 countries that had received World Bank loans was conducted to highlight their most common infrastructure problems. Financial and wage-labor problems were the leaders, followed by unclear goals and lack of management autonomy and accountability. This chapter spells out problems that make road agencies inefficient: human resource constraints, inadequate financing arrangements, lack of clearly defined responsibilities, inefficient management structures, and weak management systems.

The Institutional Framework

Part of the blame for poor road maintenance policies comes from the institutional framework within which roads are managed. They are not managed as part of the market economy with its formidable pricing dynamic. There is no clear price for roads, road expenditures are most often financed from general tax revenues, and the road agency is not subjected to any rigorous market discipline. These bias managerial incentives. Roads are managed like a social service with multiple goals. Road users pay taxes and user charges, but the proceeds are almost always treated as general tax revenues. Instead of being financed through user charges, roads are thus financed through budget allocations determined as part of the annual budgetary process. These allocations bear little relationship to underlying needs (that is, to the cost-effectiveness of road expenditures at the margin) or to users' willingness to pay. There is no hard budget constraint (that is, no direct link between revenues and expenditures), no price to ration demand (do users want more or less of particular road services?), and expenditures are not subjected to the rigorous tests of the marketplace (how much road spending can the economy afford?).

Because road users do not pay for roads directly, they are not forced to choose whether and how to make a journey, or to hold the road agency accountable for the way it spends its budget. Further, there is a free-rider effect in that the absence of a firm link between revenues and expenditures encourages individual road users to demand more road spending because it does not affect individual payments for road use. Finally, without a hard budget constraint and pressure from road users, the road agency does not have to manage resources efficiently. The government rarely provides clear objectives (road agencies are often required to employ too much labor and to build roads that are uneconomic), managers face few incentives to cut costs (major cost reductions may simply lead to reduced budget allocations in the next period), there are few sanctions, staff cannot easily be disciplined, and managers are rarely penalized for poor performance. There is thus a need to instill an efficiency ethic, a standard for productive use of all resources.

Human Resource Constraints

Human resource constraints is the most important issue facing many road agencies. They suffer from an acute shortage of technically qualified staff and at the same time employ far too many unskilled workers.[1] Morale is generally low, primarily because of low salaries that compare poorly with those in the private sector. Furthermore, the incentives working on individual managers and technical staff often discourage initiative, diminish personal accountability, and further depress morale. Rewards are not given for exceptional performance, and sanctions are not imposed on poor performers. Furthermore, staff are often inadequately trained to carry out their new professional responsibilities as road agencies change from being service providers to clients. One cannot manage a road agency with a demoralized, poorly trained, and part-time staff that has little incentive to work effectively or efficiently.

Overstaffing is still a problem for some road agencies, particularly in parts of the Middle East and North Africa, South Asia, and Latin America and the Caribbean. Having a large number of staff doing work in-house means that, in the face of ever-diminishing funds, the payroll takes up a growing share of total expenditures, leaving less available for actual works. For example, in Uruguay more than 40 percent of the national road agency's budget was spent on the payroll in 1995. In the Dominican Republic during 1994 about 3,000 staff were employed to maintain a network of 5,000 km—and salaries absorbed more than two-thirds of the total maintenance budget. In Argentina some provincial road agencies had several hundred workers per 1,000 km of road even though estimated requirements were between 50 and 80 workers. The staff of Jordan's Ministry of Public Works and Housing assigned to roads numbered about 8,000 in 1997, and this on a road network of only 7,000 km. Compare this figure with the 688 staff employed to deal with road construction and maintenance on the Ghana Highway Authority's network of 14,100 km (table 3.1).

At the same time poorly conceived or implemented retrenchment can also be counterproductive. The most skilled engineers and managers can be lost, and the remaining employees demoralized. In Kazakhstan the rapid pace of reform and arbitrary rationalization has cut the adequate pool of capable staff to barely a dozen in the Department of Roads, which is responsible for managing 17,000 km of main roads and supporting the regional management of another 70,000 km of secondary roads.

Engineers working in the private sector generally earn more than twice as much as their public sector counterparts (tables 3.1, 3.2). At one time agency staff enjoyed perks that made up for lower salaries. But inflation has eroded these fringe benefits, and private sector employers now tend to offer better bonuses, housing allowances, and car allowances. Real salaries have also declined sharply. Salaries in some road agencies are so low that "daylighting"—working another full-time job during regular working hours—has become part of the status quo; employees' primary allegiance is with other employers.[2] Some road departments even implicitly acknowledge the disparity in salaries by officially condoning daylighting as a supplement to public salaries. This has occurred in the Dominican Republic.

Road departments paying qualified technical staff a fraction of the going market wage end up with high vacancy rates, employ expatriate road managers paid through donor-financed technical assistance programs, or use part-time staff forced to supplement their incomes by moonlighting, daylighting, manipulating allowances, and pilfering.[3] This problem cannot be solved through training, bonded studentships, and improved allowances. There is no point to training staff who spend only a fraction of their time on the job. Likewise, bonded graduates have no interest in making a career in the road department and leave as soon as their bonding period ends. Improved allowances are equally ineffective since they are discretionary, subject to change, and are not bankable, that is, cannot be used as security for mortgages and other loans.

Staff in many road departments are still not held personally responsible or accountable for their work. Typically, workers do not have any job specifications to guide them, or if some do exist they do not relate well to the reality of the post. Staff do not know what is expected of them, which decisions they should make, and which decisions should be passed up or down the hierarchy. This uncertainty not only paralyzes decisionmaking but stifles initiative and detracts from job satisfaction. Moreover, managers do not know which skills are needed at particular levels and so are unable to judge whether or not workers possess them. Hence, it is impossible to assess performance fairly, allowing greater leeway for political or personally driven per-

Table 3.1 The number of technical staff and salary scales in selected countries, 1996–97

Country	Road length (km)	Number of staff	Kilometer per staff member	Annual salary range (1996–97 dollars)
Argentina	8,328			
Engineers		247	34	20,000–24,000
Technicians		800	10	14,000–20,000
Chile	46,979			
Engineers		731	64	8,436–27,000
Technicians		306	154	4,380–9,012
Ghana	15,232			
Engineers		125	122	1,500–2,000
Technicians		383	40	1,000–1,500
Hungary	30,000			
Engineers		561	54	6,315–10,500
Technicians		226	133	5,260–8,420
Jordan	7,041			
Engineers		60	117	5,400–8,000
Technicians		90	78	1,800–6,000
Kazakhstan	87,300			
Engineers		365[a]	240	2,800–4,500
Technicians		80[a]	1,090	1,800–3,000
Korea, Rep. of[b]	12,052			
Engineers		35	344	17,000–20,000
Technicians		3	4,017	12,000–16,000
Pakistan	6,580			
Engineers		294	22	2,040–5,760
Technicians		1,251[a]	5	840–2,040
South Africa	6,133			
Engineers		63[c]	97	32,000–46,400
Technicians		111[d]	112	15,600–28,000

a. Figure also includes some nontechnical administrative staff.
b. Figures for Bureau of Public Roads only. Most staff work at the provincial level for which figures are not available.
c. Includes 8 vacant positions.
d. Includes 21 vacant positions.
Source: Country national road agencies and World Bank task managers.

formance appraisal, if indeed such appraisal is made at all. In some road departments promotions are still made on the basis of administrative criteria rather than merit, and senior staff are often political appointees, with consequently limited technical capacity and autonomy. Political appointees are also likely to come and go frequently during times of political change, causing discontinuity and disrupting management of the network and staff.

A final human resource constraint is an imbalance in professional skills. Staff in road agencies tend to be engineers who may be strong in the technical aspects of building and maintaining roads, but weak in the analytical and managerial skills needed to look after a network in the long term. Standardized maintenance strategies, common in many regions, especially in the countries of the former Soviet Union, have deskilled engineers by removing professional judgment from decisionmaking.

Table 3.2 Incomes of public and private sector engineers in selected countries, 1997
(dollars per month)

	Argentina	Chile	Ghana	Hungary	Jordan	Korea, Rep. of	Pakistan	South Africa
Public salary	1,800	750	160	625	500	1,650	200	2,500
Private salary	5,000	1,750	825	1,250	750	2,800	485	2,500
Private/public	2.8	2.3	5.1	2.0	1.5	1.7	2.4	1.0

Note: Figures are of salaries and allowances for graduate engineers with three to four years practical experience.
Source: Country national road agencies and World Bank task managers.

In much of the world the role of government is changing in all sectors, including roads. Yet this move is often not accompanied by training for professional staff to cover such key areas as planning and economic analysis, environmental assessment, contract management and supervision, and prioritization of works. Likewise, engineering skills are not being updated to include modern construction and maintenance techniques.

Inadequate Financing Arrangements

All countries suffer from a shortage of funds for roads, a shortage of funds for both investment and maintenance. Low investment results in high congestion, often intensified by the frequency of lane closures because of the need to repair deteriorating pavement and structures. We see this problem especially in rapidly growing cities, such as Bangkok, Buenos Aires, Manila, and Mumbai. Lack of funds for maintenance results in the decay of road networks (see World Bank 1998). The initial impact of this funding crisis has been to increase road transport costs, in terms of travel time; VOCs; road conservation; pollution; and road accidents. The long-term impact has been to reduce commercial and agricultural competitiveness in international and regional markets and consequently slow overall economic growth.

Without an adequate and stable flow of funds, road maintenance policies will not be sustainable. Maintenance expenditures in virtually all countries are well below the levels needed to keep road networks in stable condition for the long term. In many countries these expenditures are less than half the amount required and, in some, less than a third (table 3.3). And this problem is not confined to only transition or developing economies. Many of the wealthiest nations in the world are also failing to properly finance road maintenance. For example, in Canada spending on public roads in 1993 relative to road traffic volumes was only half that of 1965. In the United Kingdom a 1996 Institution of Civil Engineers survey of local authority networks found that maintenance of local roads (96 percent of the total road network) was being underfunded by $1,440 million per year, and construction and improvement by a further $2,260 million.

What's more, the flow of funds is erratic. Budget allocations are often cut at short notice in response to difficult fiscal conditions, funds are rarely released on time, and actual expenditures are often well below agreed allocations. As a result road agencies are unable to plan works effectively, contractors are not paid on time and go out of business, short-term "patch and mend" work replaces appropriate road conservation, rural roads regularly become impassable during the rainy season, and the large backlog of road rehabilitation continues to grow. Between one-quarter and one-third of the main and secondary road networks included in table 3.3 are in poor condition and must either be rehabilitated or downgraded to roads that receive minimal maintenance.

Too Few and Poorly Allocated Funds

Road maintenance is underfunded mainly because road users do not pay enough for their use of the road network (see table 3.3). They pay the usual import duties and excise and sales taxes—but so does everyone else. Since private cars are a luxury good for low- and middle-income economies, a higher level of general taxation, at least on private car ownership and use, would be justified on equity grounds. Yet road-user charges—in the form of vehicle license fees, a specific surcharge added to the price of fuel (the fuel levy), and international transit fees—rarely cover more than 50 percent of expenditures on road maintenance and, in some countries, barely 25 percent (see box 3.1).

Most road expenditures are still financed from general tax revenues (listed in table 3.3 as "government grants") and donor-financed loans and grants. Moreover, the underpricing of road use has led to the dramatic shift to road transport worldwide, primarily at the expense of railways. Demand has not been managed through appropriate pricing. But it need not be. Roads can be commercialized, put on a fee-for-service basis so that demand can be better managed, and treated like any other public enterprise.

An added complication is that funds for road maintenance are allocated as part of the annual budgetary process. But in the absence of proper network assessments financial proposals for maintenance are based on historical spending patterns rather than real need. Each ministry must compete for funds during the annual budget negotiations.

Table 3.3 Main and secondary road expenditures, financing, and actual and required maintenance in selected countries

(millions of dollars)

	Argentina 1995	Chile 1995	Ghana 1996	Hungary 1995	Jordan 1994[a]	Kazakhstan 1996	Korea, Rep. of 1997	Pakistan 1995	South Africa 1995
Road expenditures[b]	333	526	191	348	74	101	5,768	395	874
Amount financed by									
Road users[c]	0	105	50	205	0	20	5,593	1	0
Government grants	293	421	20	97	58	81[d]	175	320	874
Donors	40	0	121	46	16	0	0	75	0
Maintenance expenditures									
Required[e]	68	760	135	240	31	176	970	40	742
Actual	61	308	73	127	9	101	655	27	507
Maintenance shortfall	7	452	62	113	22	75	314	13	235
Actual/required (percent)	89	41	54	53	29	57	67	68	68

a. Jordan figures are from the Budget Law of 1996.
b. These figures represent actual spending, which is generally below requirements, due to shortfalls in regular road maintenance.
c. Includes license fees, international transit fees, and fuel levies where directly channeled to finance roads.
d. All from a turnover tax (not a road user charge) dedicated to the road fund.
e. Maintenance requirements from country road agency estimates. Most include some element of rehabilitation.
Source: Survey of country road agencies, World Bank sector and project reports, and World Bank task managers.

In theory, funds are allocated to finance those expenditures with the highest economic return, which would ensure that road maintenance would not be underfunded. But in fact allocations for maintenance are well below the optimal requirements (defined as a maintenance strategy that produces an economic internal rate of return of more than 12 percent), even though the economic return at the margin is frequently more than 100 percent.

The budget allocation process is flawed and politicized, and large spending ministries, particularly those proposing to spend high sums on maintenance, nearly always lose out in budget debates. Lack of funds for maintenance does not lead to immediate, catastrophic failure and there is thus little political pressure or incentive to support maintenance. Likewise, maintenance can always be postponed in the hope that better fiscal times are around the corner. But they rarely are, and road maintenance continues to be cut or deferred. Given this inherent structural problem, it is no wonder that some industrial economies have turned to earmarking to secure a stable flow of funds for their road expenditure programs (box 3.2).

Too Much New Investment

Road maintenance is also underfunded because some countries still spend too much on new investment (mainly upgrading existing roads and building feeder roads)—scarce resources are misallocated. Lack of market discipline has encouraged governments to minimize their own road maintenance expenditures, disregarding the impact that this has on total road transport costs. Further, maintenance is normally financed under the recurrent budget, and recurrent revenues are nearly always in short supply. Since in the past donors have been willing to finance rehabilitation under the development budget (often on a grant basis), governments have had every incentive to capitalize road maintenance and charge it against the development budget. Rehabilitation, rather than recurrent maintenance, became the "optimal" solution.

Donors have since recognized this mistake and most will no longer finance rehabilitation programs until governments have introduced sustainable road maintenance policies. But a further and perhaps more important reason for favoring new construction is that such contracts tend to be larger (hence offering greater opportunities for gratification payments) and are politically more visible and glamorous.

Other countries have no choice but to invest heavily in new construction as there are many potential construction projects with a high economic internal rate of return and demand for road space outstrips supply to such an extent that economic growth is severely impeded. In many economies in East Asia and, to a lesser extent, Latin America, the costs of inadequate

---

Box 3.1 Separating road-user charges from general tax revenues

The taxes and charges paid by road users are generally identifiable as: specific charges for use of the road network (for example, tolls, fuel levies paid into a road fund, and vehicle license fees); "green" taxes imposed on road users to try and internalize the external costs of road use; general revenue taxes (for example, value added taxes, corporate income taxes, and trade protection taxes); or taxes used to collect road-user charges and raise general revenues (generally only excise taxes, but may also include some import duties and sales taxes). Since it is fairly easy to identify the taxes and charges that fall into the first three categories, this box concentrates on ways of dealing with road-user charges and general taxes.

When road user charges are combined with other general taxes, they add to the existing indirect taxes (for example, taxes on goods and services and import duties). Indirect taxes generally differentiate among consumer luxuries, other consumer goods, intermediate goods (including raw materials), and capital goods. Within each category items are usually treated in a fairly consistent way, although there are exceptions since tax rates also reflect other fiscal objectives (such as promoting domestic vehicle assembly, energy conservation, and protection of local industry). The following four-step procedure is suggested as a means to separate road user charges from general revenue taxes.

• First, prior information will often be available to show how the overall tax rate has been built up and how much of the overall rate comprises the road-user charge. For example, in China the purchase tax on new vehicles includes an added vehicle purchase fee, which is credited to a special fund to support road construction. Unfortunately, information on the structure of the tax rate is not systematically recorded and may not be readily available. But when it is available, it may enable the road user charges to be separated from general revenue taxes.

• Second, when there is no prior information, it is worth examining the tax code to see how the taxes levied on road users compare with the taxes levied on other goods and services. For example, trucks are usually classified as plant and equipment. If the tax schedule levies the same rate on trucks as on all other plant and equipment, then, prima facie, there is no road-user charge added to the tax rate. On the other hand, if the rate is clearly higher than that on other plant and equipment, the difference may represent a road-user charge (the difference would represent the maximum amount that could be considered a road user charge since the additional element may reflect other fiscal objectives).

• Third, when it is not possible to identify the tax rate applicable to road users, the analysis must rely on the average tax rate for all similar goods. For example, the rates applicable to individual items of plant and equipment may vary widely, and in such cases there may be no alternative but to use the average rate as representative. The average is calculated by dividing the tax revenue collected from a particular tax (such as general sales taxes, excise taxes, and import duties on plant and equipment) by the base value of these items. The difference between the taxes levied on road users and the average tax rate on the group as a whole may then be treated as the road user charge (again, this amount represents the maximum that can be considered a road user charge). This procedure is not particularly satisfactory and should be avoided if possible.

A recent unpublished study applied the above method to eight countries (Argentina, Bangladesh, Bolivia, China, Indonesia, Mexico, Tanzania, and Turkey) and showed that import duties, sales taxes, and excise taxes rarely include an additional element representing a road-user charge. Indirect taxes are nearly always general revenue taxes and neither charge directly for use of the road network or raise revenues specifically for roads.

---

road infrastructure are large, as are the corresponding investment needs. A 1992 study by the Korea Transport Institute estimated that the costs of congestion on intercity and urban routes was $6 billion per year, or about 2.5 percent of GNP. Hence, not only is investment in the transport sector projected to be a high percentage of GDP (5.2 percent in 1993–97), but roads make up the majority of total investment (56 percent). In China the projected need for new roads is also large: $504 billion over 1996–2010. Likewise, the OECD countries with the highest growth rates in car and truck traffic—including Germany, Japan, Portugal,

Turkey, and the United Kingdom, all with traffic growth rates higher than 3 percent per year—are spending more than 50 percent of their total road budgets on new construction (OECD 1994).

Lack of Clear Responsibilities

A lack of clearly defined responsibilities adds to the above problems. It is often not established which agency is responsible for managing different parts of the road network, controlling overloading, managing

Box 3.2 Earmarking and user pay in Japan, New Zealand, and the United States

Several countries responded to the rapid expansion in demand for road transport in the post-war period by establishing road funds. The concept of "user pay" stood behind the establishment of such funds—the road user pays certain road-related taxes and the government credits the proceeds directly to a special highway account.

JAPAN ROAD IMPROVEMENT SPECIAL ACCOUNT. This special funding system was introduced in 1954 to meet the needs of the post-war road improvement program. It was "based on the concept that road users who enjoy the benefits of improved roads should bear the burden for their improvement." It includes an elaborate system for earmarking national and local taxes, both supplemented by general revenues, to finance the maintenance, improvement, and construction of roads.

At the national level earmarked tax revenues consist of 25 percent of gasoline tax revenues ($0.39 per liter), half of tax revenues on liquid petroleum gas ($0.14 per kg), and 75 percent of revenues from the motor vehicle tonnage tax ($51 per 500 kg per year). At the local level earmarked tax revenues consist of: tax revenues collected by the national government and then passed on to the local government (the other half of the liquid petroleum gas tax, a local gasoline tax of $0.04 per liter, and 25 percent of the motor vehicle tonnage tax) and tax revenues collected by the local government itself (a local diesel fuel tax of $0.26 per liter and the motor vehicle purchase tax currently set at 5 percent of the purchase price). Revenues from all these sources amounted to roughly $30 billion in 1995.

The tax rates are set during the preparation of the Five-Year Road Improvement Programs. The Ministry of Construction prepares the Programs in consultation with local governments and then submits them to the Ministry of Finance for approval. After discussing the proposals, new tax rates are agreed on and written into a new proper tax law, which remains in force for the next five years.

U.S. FEDERAL HIGHWAY TRUST FUND. The Trust Fund was introduced in 1956 to finance construction of the interstate highway network. The Fund is based on the user-pay concept, which is well established in the United States. All but six states now dedicate their state-level user-fee revenues to special highway or transportation accounts.

The Trust Fund revenues derive from a variety of highway user taxes, including motor fuel taxes on gasoline, diesel, and gasohol (currently $0.14, $0.20 and $0.08 per gallon, respectively); a graduated tax on tires weighing 40 pounds or more; a 12 percent retail tax on selected new trucks and trailers; and a heavy-vehicle use tax on all trucks with a gross vehicle weight more than 55,000 pounds. Total revenue raised through these taxes was $21 billion in 1995, most coming from the tax on gasoline. Tax rates are adjusted as part of the regular budgetary process.

Revenues from the highway portion of the Trust Fund are used to reimburse states, on a cost-share basis, for expenditures on approved projects. These include periodic maintenance, road improvement, new construction, road safety, road studies, and other highway-related expenditures, except for routine maintenance. Since 1982 a portion of the Fund has also been used to finance mass transit projects, and, since 1991, its mandate has been extended to supporting other land transport modes.

NEW ZEALAND NATIONAL ROADS FUND (NRF). The original road fund was established in 1953, although the latest version, the National Roads Fund, was created in 1996. The fund derives revenues from a fuel excise added to the price of gasoline (currently $0.065 per liter); weight-distance charges on diesel vehicles purchased as distance licenses and approximately proportional to gross wheel-load (and hence more closely related to damage imposed on the road pavement); and motor vehicle registration and license fees. The charges raised $589 million in total—$195, $286 and $108 million, respectively—in 1996–97. About $30 million per year, that is, 5 percent of revenues, is spent on collecting the various user charges under contract.

The government still sets the level of charges on the recommendation of the Treasury, Ministry of Transport, and the agency responsible for controlling the funding of roads, Transfund New Zealand, and hence determines the inflow of resources. But it no longer determines the outflow. Once the costs of policing the road network and the Land Transport Safety Authority have been met ($93 and $15 million per year, respectively, in 1996–97), Transfund can use the balance of the revenues without further interference.

Transfund finances the entire cost of the national road agency, Transit, as well as regional planning activities and local authority networks on a cost-sharing basis. Bids from all road agencies are subjected to a benefit-cost analysis before prioritization, with a cutoff at a ratio of 4 or less. All bids for maintenance works have to be based on a standardized maintenance management system.

urban traffic, improving road safety, or reducing the adverse environmental impacts associated with road traffic. The poor definition and enforcement of responsibilities at the individual staff level also relates here.

Responsibility for roads is often diffused among several central government ministries and local government agencies, leading to duplication, confusion, and a lack of coherent management policies. For example, in Jordan six organizations—three ministries, one municipal government (the Municipality of Greater Amman), and two parastatals. are responsible for various parts of the road network The situation is further complicated in that the Ministry of Public Works and Housing is responsible not only for all primary and secondary roads, but also for some village and agricultural roads. In Indonesia the principle responsibilities for the road network are divided between two ministries (Public Works and Communications), while another four agencies and ministries are also involved in road transport. This multiplicity of participating agencies is not limited to the developing world. In the United Kingdom the administration of roads involves the Department of Transport, Environment and the Regions; the Highways Agency; regional government offices; and local authorities. In many countries traffic regulation and enforcement are handled by a separate transport ministry and the police, further complicating matters.

In addition, different road agencies rarely have distinct responsibilities. For example, it is often uncertain whether the main road agency or the urban municipality are responsible for trunk roads in urban areas. A relatively common scenario is that of an urban through-route constructed or upgraded by the national road agency, often through a loan or grant, but with responsibility for maintenance left uncertain. Responsibilities between national and regional road departments are also often not clear cut. Frequently, decisionmaking is too centralized—decisions that should be made by staff in regional offices who know what is happening in the field are instead made in the main roads department in the capital. Moreover, road classification systems are often out of date. New roads may not have been listed, and changes in the functional class of existing roads may not have been accompanied by the appropriate reassignment of responsibilities.

This problem is even more acute in rural areas. Rural roads in many countries have never been formally assigned to a legally constituted highway authority. Some have been built using central government funds or multilateral and bilateral donor grants channeled through central government departments dealing with agriculture, fisheries, and tourism. In several countries in Africa and Asia no arrangements were made for transferring managerial responsibility for these roads to an established road agency. National and local road agencies did not know which roads they were supposed to maintain, and many rural roads went unclaimed and unmaintained. For example, in Russia about 450,000 km of enterprise roads are undesignated, and a further 700,000 km of access roads have no legal owners. While there is often local pressure to transfer the enterprise roads to regional authorities, these authorities are understandably reluctant to take on the burden of additional maintenance funding, since most of the roads do not meet public road design standards.

Most road agencies are also unclear about their responsibilities for a number of other important road traffic activities. Among these is axle weight enforcement. Regulations are commonly promulgated by the transport ministry, and administration and enforcement are handled by a number of agencies, including road departments, traffic commissioners, the police, and private contractors. Reviews of axle weight enforcement have identified several key weaknesses: poorly assigned responsibilities, weak enforcement agencies, and resistance by truck owners and operators. The road agency frequently has no incentive to enforce regulations or to prosecute offenders, as any fines (which are anyway too low to act as a deterrent) typically accrue to the consolidated account, while the costs of enforcement are charged against the road agency's budget.[4]

Main road agencies are sometimes unclear as to whether they should actively intervene to manage urban traffic by enforcing parking and other traffic regulations. Often this job is left to municipal governments. This ambiguity is largely a result of confusion over the allocation of responsibilities between central and local governments on urban through-routes and is closely linked to the construction of new relief roads and bypasses or the improvement of main roads that cross urban centers. Assessing the environmental impact of new road schemes is also an area of increasing concern, as is assigning responsibility for identifying and mitigating the adverse impacts associated with roads and road traffic. Finally, there are ambiguities surrounding less important issues, such as liability claims for accidents caused by defective design and maintenance policies, as well as compensation from third parties for damage done to road infrastructure by road accidents and utility companies.[5]

Many of these road and traffic-related problems are aggravated by a shortage of technical staff and underdeveloped legal and administrative systems. But the core problem is lack of clearly defined responsibilities—most often caused by the absence of a coherent legal framework and cogent mission statements for the various road agencies.

Ineffective Management Structures

The problems discussed above are worsened by the diverse management structures under which most roads are administered. The central government usually manages the main road network in one of four ways:

• As part of a combined ministry of works, transport, and communications, such as in Hungary, the Netherlands, Sri Lanka, and Tanzania.
• As part of a more narrowly focused ministry of works or transport, such as in Chile, Indonesia, Jordan, the Philippines, and Zambia (figure 3.1).
• Under a sharply focused ministry of roads and highways, such as in pre-1997 Ghana.
• As an arm's-length road agency reporting to any type of parent ministry, such as in Argentina, Ghana, Latvia, and the United Kingdom.

The model illustrated in figure 3.1 is cumbersome and, in practice, largely ineffective as a framework for promoting a more commercial approach to road management. Regional engineers often report directly to the permanent secretary instead of through the director of roads, numerous support services are shared (and suffer from conflicting priorities), and the structure is lopsided. While the road sector has grown rapidly relative to other sectors, this increased importance has not been reflected in changed priorities or the changed relative status of departments.

The management structures of most road agencies date back to the time when the ministry of public works spent about as much time on roads as it did on maintaining public buildings and procuring government vehicles. Times have changed. Today road departments typically account for more than 70 percent of the ministry's total expenditures and manage more assets than either the railways or the national airline. Nevertheless, the head of the road department is usually appointed to a level corresponding to that of the chief civil engineer in the railways or chief mechanical engineer in an airline. Furthermore, the organization of a typical central government road agency exhibits three structural weaknesses: it is missing a layer of management, it is usually overly centralized, and the director of roads rarely reports directly to the permanent secretary.

A more focused ministry would overcome some of the above problems since reporting lines would be more direct. Furthermore, a more narrowly focused transport ministry would benefit from better intermodal coordination, although the ministry would still remain lopsided, the management structure weak, and the director of roads still a line manager. The special-purpose ministry, such as that in Ghana before 1997, provides the simplest model, although the same objective could be achieved by restructuring a larger, heterogeneous ministry.

Urban and rural roads may be handled directly through a central road agency (such as in Sierra Leone), through a separate department forming part of a central road ministry (such as the Local Government Engineering Department in Bangladesh), or, more commonly, by local authorities. Local authorities sometimes work through a local government ministry, which in turn usually delegates most day-to-day operations back to the local authorities.

At the level of local government, management structures tend to be even more confused. There is often no road department (new roads are typically the responsibility of the development committee and road maintenance the responsibility of the finance committee) making it difficult to identify who is responsible for what. Usually, staff are demoralized, underpaid, inexperienced, poorly skilled, and unmotivated. Local government entities are typically small and lack both the

Figure 3.1 Typical management structure of a ministry of works and transport

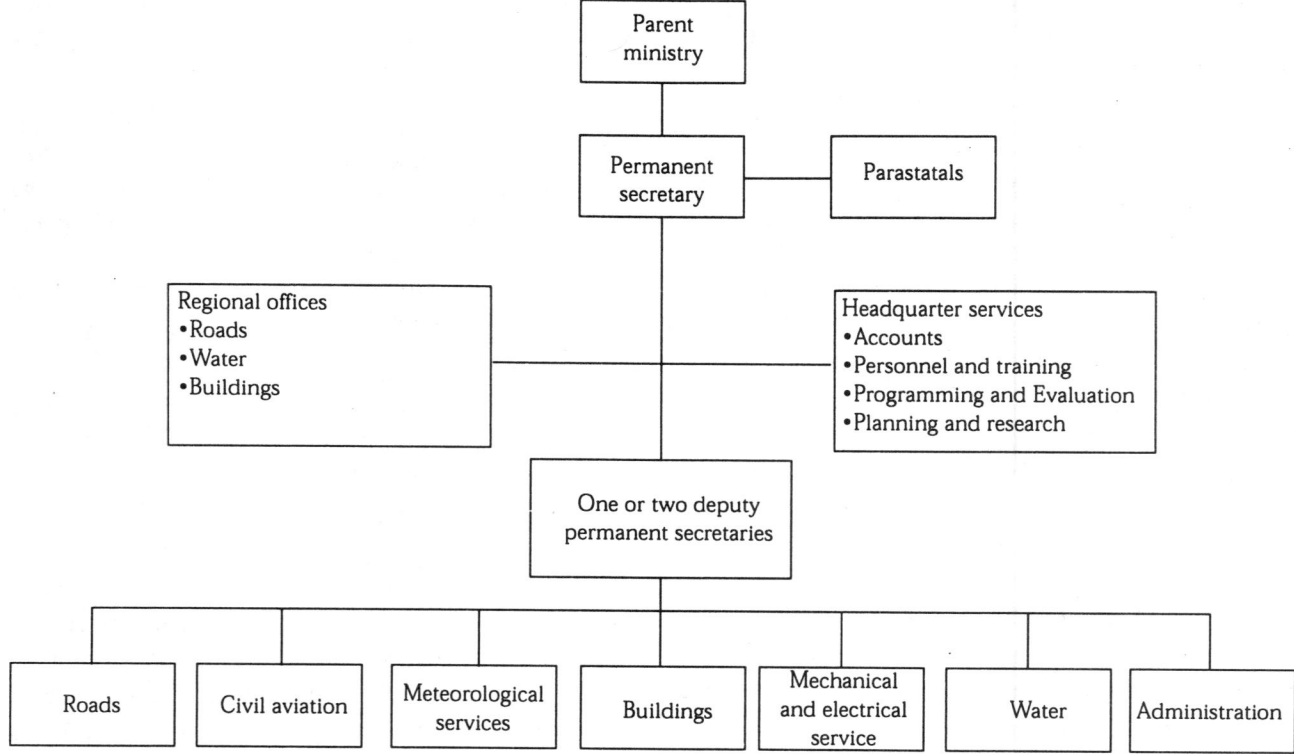

technical capacity and the resources to manage their road networks effectively. These problems are matters of scale: local governments are too small to justify hiring people with the skills needed to plan and manage road networks acceptably.

Weak Management Systems

Effective management requires timely collection and analysis of both physical and financial information. Yet many road departments do not possess even the most rudimentary management information systems. Moreover, confusion and poor management structures offer managers little incentive to introduce and develop such systems.

Financial accounting systems often provide little information to support enlightened management decisions. Typically, there is no revenue account (hence no cash flow statement), accounts are kept on a cash basis, and investments are written off as a cash expense as soon as they are incurred (that is, road agencies do not keep a balance sheet or depreciate assets). Accounting

systems often use line-item budgeting with very broad cost headings that involve a great deal of aggregation. Items like "administration," "rent," and "electrical and mechanical" frequently cover several functions, and there is no simple way to identify the expenditures attributable specifically to roads. Most road departments cannot tell how much they spend on routine and periodic maintenance, since some periodic maintenance costs are charged to the recurrent budget and some to the capital budget. They cannot discern the breakdown of costs among overhead, labor, and equipment, or the unit costs of shoulder repairs, regraveling, and cleaning drains. Such poor accounting systems make it difficult, if not impossible, for managers to establish consistent spending priorities.

Numerous attempts have been made to introduce management information systems, but with little success (box 3.3). Many fail as soon as the consultants who have installed them leave (box 3.4). The most recent World Bank review of road management systems showed that basic roads inventory data were valid or complete in only 10–25 percent of countries in Africa, Latin America and the Caribbean, and Asia, and 50 percent in the Middle

---

Box 3.3 Problems implementing road management systems

Robinson and May (1997) reviewed donor-financed projects in which consultants were employed to develop road management systems. The aim was to identify why these projects encountered problems during implementation and what might be done to avoid such problems in future. The following external, institutional, and technical factors were identified as important in contributing to poor performance:

External
Economic and financial problems:
• Weak local economies and foreign exchange shortages preventing the purchase of required inputs.
• Local funds sufficient to cover only staff salaries, leaving little to finance actual maintenance.
Cultural problems:
• Problems introducing modern management practices, including incentives, into cultures that were ruled by elders, suffered from ethnic problems, or where nepotism and favoritism prevailed.
• Traditional behavior in which excuses had to be made to avoid blaming individuals.

Institutional
Client attitudes:

• Lack of commitment, often because the road management system had been imposed by donors as a loan condition.
• Expectation of high-tech solutions when simple common-sense solutions were more appropriate.
• Greater than expected resistance to change.
Design problems:
• Inappropriate and unrealistic terms of reference.
• Consultants lacking sufficient qualified staff.
• Systems too complicated to be sustained by local staff and domestic resources.
Staffing and training problems:
• Shortage of experienced local staff.
• Operational requirements preventing local staff from being released for training.
• Overambitious training programs with poorly prepared instructors.
• Insufficient follow-up training and updating.

Technical
Hardware and software problems:
• Deficient computer facilities and unsuitable hardware.
• Inadequate data.
• Focus on procuring new equipment, rather than the systems needed for maintenance and repair.

Source: Robinson and May (1997).

East and North Africa (the latter region included Hungary, Portugal, Turkey, and the former Yugoslavia). Data on pavement condition, surface roughness, and pavement strength were virtually nonexistent in 30–50 percent of the countries surveyed in Africa and Asia.[6] Timely and accurate traffic counts, essential for informed road planning, are often incomplete, with road agencies conducting surveys on unrepresentative, irregular, or nonexistent lengths of the road network.

Although donor-financed lending programs over the past two decades have typically included maintenance management systems, only 10 percent of the countries surveyed in Africa, and 30–50 percent elsewhere, had a functioning routine maintenance management system. Moreover, performance was variable. Likewise, economic evaluations of road maintenance interventions were still uncommon. Although the vast majority of World Bank projects use the World Bank's Highway Design and Maintenance Model, only 15 percent of the countries reviewed used it—or a similar economic evaluation model—to program regular maintenance. The use of bridge management systems was even less common.

Many countries do not have functioning maintenance management systems to determine network-wide maintenance priorities. Fewer, still, supplement such physical planning tools with performance budgeting systems. A country cannot manage a large road network efficiently without a usable management information system to help managers set priorities and monitor performance against predetermined targets. Yet this is precisely what many road agencies in developing and transition economies are still trying to do.

Inefficient Work Methods

Few road agencies manage their resources well enough to achieve reasonable value for money. Instead, they

---

Box 3.4 The failure of a maintenance management system

A 1995 consultant's report had this to say about a failed maintenance management system in a Middle Eastern country: "The Road Maintenance Management System (RMMS) was introduced in 1986 when a complete inventory and inspection of the network in each district was prepared. It was intended for this to be updated each year. The RMMS was to be used to determine the allocation of funds for maintenance between districts, although it was not suitable for determining rehabilitation and reconstruction priorities. However, the Directorate of Road Maintenance in the Ministry of Public Works has not used the system since 1992, and the RMMS is now in complete disarray and disrepute for no clear reason. The following reasons have been offered by various officials involved in the program, all of which are valid to a greater or lesser degree:

- some districts do not make returns;
- district officials do not have the staff, vehicles and

equipment to update the inventory and inspect the roads;
- maintenance engineers in the districts do not understand the condition categories due to lack of training;
- the labor force in each district is too large, most are unwilling/incapable of producing reasonable quality work, but it is impolitic to discontinue their employment;
- there is great difficulty in obtaining sufficiently skilled and responsible supervisory staff; and
- the financial allocation is only sufficient to pay the wage bill.

Although there is nothing fundamentally wrong with the system, it seems to have failed through inadequate management, lack of training, ill-designed personnel policies and underfunding. It seems probable that the system has been seriously undermined due to political and other pressures on district engineers to employ more daily laborers than are needed and without regard to their capacity for work."

Source: Unpublished consultant study by Halcrow, Fox, and Associates.

---

deliver poor quality services—the result of meager annual budgets—characterized by undue reliance on work done using in-house staff and equipment, inefficient operation of government plant pools, and lack of interest in labor-based work methods.[7] These practices are typical of agencies that face no market discipline and have poorly motivated and unaccountable managers.

A considerable portion of maintenance, particularly routine maintenance, is still carried out using in-house staff and equipment, even though its quality is variable and costs usually higher.[8] Although cost comparisons are often tenuous, in-house work exposed to private sector competition nearly always dramatically increases efficiency, with costs falling by as much as 30 percent.

Contract maintenance can also improve quality. A recent review of experience with contract maintenance in six Latin American countries concluded that such practices had helped to solve, or at least alleviate, inefficient resource use (World Bank 1996c). Still, contract maintenance is not a panacea. Road agencies are naturally reluctant to give up the power that comes from managing large in-house labor units. And, further, contracting out will work effectively only when procurement procedures are straightforward, that is, if there is a healthy and competitive local construction industry and a stable flow of funds to pay the contractors. The road

agency must also have enough qualified staff to process contracts, supervise work, and deal with arbitration issues—tasks for which staff in many countries, like those of the former Soviet Union, often lack experience.

Inefficient government plant pools are another symptom of weak market discipline. Most road agencies own millions of dollars worth of heavy plant and equipment, much of it procured under loans from international donors or furnished on a grant basis by bilateral donors. Even a relatively small road agency may own plant and equipment worth $50 million or more. Utilization rates for this equipment often drop to 20–30 percent, compared with 80–90 percent in the private sector. The economic losses associated with these low utilization rates can amount to more than $23 million per year.[9]

The superficial reasons for such low utilization rates include poor management and accounting systems, lack of standardization, shortages of fuel and spare parts (or shortage of foreign exchange to purchase them), shortages of trained equipment operators and mechanics (mainly due to poor terms and conditions of employment), and overstaffing of unskilled laborers. The real reasons are related to lack of a stable work load (that is, inadequate road maintenance allocations and an erratic flow of funds), lack of transparent management systems (that is, costing systems that clearly spell out the price of low utilization levels), lack of managerial accountability, and political interference. No one knows, or cares, that allowing equipment to be idle involves serious waste.

The lack of interest in labor-based work methods is also symptomatic of weak market discipline. Not only are labor-based methods often much cheaper (in Tanzania and Ghana labor-based contracts cost about 30 percent less than traditional contract prices), they are often more reliable because government plant pools are in such disarray. Labor-based work methods nevertheless raise other difficulties.[10] Government procurement procedures often discourage the letting of small contracts, particularly to one-person contractors who cannot be expected to follow standard bidding procedures. Donor policies, with their emphasis on international competitive bidding and preference for financing foreign exchange expenditures, add to this bias.

Evidence from a recent survey in Ghana points to two principle problems there (Stock 1996). First, large contractors have less incentive than small firms to employ labor-intensive work methods, since they do not want their costly plant to stand idle. Second, frequent delays in payments are better handled by contractors with relatively small wage bills, since payments for other inputs can more easily be deferred. Delays in paying wages lead quickly to strikes. Hence, the expectation of delays in payment for works dissuades prospective contractors from using labor-based methods. And there are other reasons. Labor-based work methods offer less scope for gratification payments, and management is under no direct pressure to find the cheapest and most cost-effective way of getting the work done.

Road agencies are unlikely to operate efficiently until they are subjected to some form of market discipline. Competition is the primary factor motivating managers to cut waste, improve operational performance, and allocate resources efficiently (Shirley 1989). But unintended and unwanted consequences could arise. If the lower unit costs achieved through improved work methods lead to a lower budgetary allocation the following year (or at least the anticipation of a lower allocation), then managers have little incentive to introduce more efficient operations.

Notes

1. Over the past few decades many governments systematically expanded the civil service, often by a factor of two or three, to deliver on election promises to reduce unemployment. Road agencies were a key target for employment programs, and some now have two to three times the required number of laborers on their books.

2. A further complication is that agency staff may work for private sector engineering firms that could compete for contracts tendered by the road department.

3. "We want them to come to work. We want them to work five days a week. We want them to work 40 hours a week. We don't want them to have to do something else in order to survive and we want them to keep their hand in their own pocket." Comments by E.V. K. Jaycox, Vice President, Africa Region, at a conference on "Capacity Building: The Missing Link in African Development," Reston, Virginia, May 20, 1993.

4. Inadequate fines not only encourage overloading and fail to cover the extra costs of the road damage, but they also present greater scope for gratification payments, thereby eroding good governance. Gratification payments can add up to substantial sums. A recent article in *The Economist* (April 12, 1997) reported the $1.25 billion annual turnover in illegal payments made by truckers to officials in order to enter Delhi.

5. Civil liability claims become more of a problem as the state of the road network deteriorates. The 1996 Local Transport Survey conducted by the Institution of Civil Engineers in the United Kingdom found that more than 70 percent of local authorities had experienced a rise in highway liability claims over the previous five years.

6. See World Bank (1991a). Most Bank-supported highway projects now include elements to redress this problem. Although a more up-to-date review might show a general improvement, the lack of an adequate road inventory and condition survey still hampers the efforts of many developing countries to manage their networks more effectively.

7. Labor-based work methods involve substituting labor for equipment in low-wage countries. See Stock and de Veen (1996).

8. Higher unit costs are, of course, hidden by the budgeting framework that makes unit cost accounting difficult, if not impossible.

9. This calculation is based on a plant pool worth $50 million, with an average utilization rate of 25 percent instead of 85 percent. The equipment is depreciated over eight years, using straight line depreciation, a 12 percent interest rate, and maximum utilization of 1,250 hours per year.

10. Accurate cost comparisons are difficult to make since they are dependent on the costing system used, market conditions, and the government's reputation as a reliable payer. Contractors often add a surcharge to contract prices to cover expected late payments.

4. Commercializing Roads: The Four Basic Building Blocks

What can be done to improve road financing and road maintenance policies? More generally, what can be done to strengthen the management and financing of roads overall? The key concept behind the reform agenda is commercialization: bring roads into the market place, put them on a fee-for-service basis, and manage them as a business.[1] But since roads are a public monopoly and their ownership is likely to remain in government hands for some time, commercialization requires complementary reforms in four other important areas, referred to as the four basic building blocks. They focus on:

- Establishing *responsibility* for managing roads by clearly assigning roles.
- Creating *ownership* of roads by involving users of roads in their management to encourage better management and to win public support for more road funding, while constraining road spending to what is affordable.
- Stabilizing road *finance* by securing an adequate, continual flow of funds.
- Strengthening *management* of roads by introducing sound business practices and enforcing managerial accountability.

The four basic building blocks are the core of reform. They are interdependent and should be implemented together. If not, reform will be only partly successful. The management and financing of roads cannot be reformed without establishing who is responsible for what. The financing problem cannot be solved without the strong support of road users. The support of road users cannot be won without taking steps to ensure that resources are used efficiently. And resource use cannot be improved without controlling monopoly power, constraining road spending, and enforcing managerial accountability. Still, there is

scope for flexibility. The reforms can be introduced in different ways, that is, the content of each building block can differ, depending on country circumstances. They can move sequentially or in parallel, and both the sequencing and the pace of reform can vary. But in the end all four building blocks should be in place to ensure that the reform agenda is sustainable and does not drift back to the status quo ante.

Assigning Responsibility

The first building block concentrates on creating a coherent organizational structure for managing different parts of the road network. This requires clearly assigning responsibility among different government departments, different levels of government, and individual road agencies. The arrangement must be based on an accurate road inventory, functional classification of roads, designation of appropriate road agencies, formal assignment of responsibility to each road agency, and clarification of the relationship between the road agency and the owner or parent ministry. Responsibilities to be assigned include those for operating, maintaining, improving, and developing the road network; for traffic management; for handling general accidents and incidents; for road accidents caused by the road agency's own negligence; and for adverse environmental impacts associated with roads and road traffic.

Creating Ownership

The second building block is concerned with the concept of ownership—building constituencies with a

32

strong vested interest in sound road management. Major policy reforms in the road sector cannot succeed without the active support of a large and vocal constituency willing to argue for better road management and additional, affordable road financing.

The obvious constituents are the stakeholders: road users themselves, together with the business community, farmers, and other people dependent on a well-functioning road network. Given that current financial allocations for roads are erratic and well below the levels needed to keep the road network in stable condition over the long term, strong stakeholder support for more road funding must be built up if reform is to succeed. The usual mechanism for winning their support is by involving them in road management. Stakeholders agree to work in partnership with the government to strengthen road management and financing in return for a seat at the table where decisions are made about how roads are to be managed and how funds are to be spent.

Ensuring Secure and Stable Financing

The third building block concentrates on establishing an adequate and stable flow of funds. Without both, none of the above reforms will be sustainable. Yet practically all governments in developing and transition economies are seriously short of fiscal revenues. Budget allocations for road maintenance rarely exceed 50 percent of requirements, and agreed allocations are often cut with little notice in response to short-term fiscal crises. Funds for road improvement are likewise in extremely short supply, particularly in rapidly growing economies and those in need of extensive road modernization.

Given these fiscal conditions, governments cannot meet financing needs by allocating additional revenues from the consolidated fund. Additional funds must come from heightened revenue mobilization. But if road-user charges are raised, there is no guarantee that the additional revenues will be allocated to roads, nor that they will generate a stable flow of funds.

Furthermore, traditional earmarking is not a viable solution. It adversely affects management of the government's overall budget and is rarely sustainable.

An added concern is that existing financing mechanisms do little to strengthen market discipline either by managing demand or by improving the efficiency of the road agency. Solving the financing problem calls for a radically new approach. Hence, the concept of commercialization. With strong stakeholder support roads can be put on a fee-for-service basis to generate the added revenues needed to support operation, maintenance, and improvement and to separate road financing from the vagaries of the government's budget.

Introducing Sound Business Practices

The fourth building block focuses on creating a commercially oriented road agency. Road users involved in the management of roads generally press for the introduction of sound business practices to ensure that their constituents get value-for-money from road spending. Road users expect clear management objectives, an effective management structure, competitive terms and conditions of employment, consolidated budgets, commercial costing systems, and effective management information systems. Introducing sound business practices changes managerial incentives. It brings pressure to dispose of in-house plant and equipment (or to use it more efficiently), to arrange for more work to be done under contract, to control vehicle overloading, and to improve road safety. These issues have become systemic sources of inefficiency in the road sector because current management procedures in most countries provide little incentive to do anything about them.

Note

1. In the early 1950s when the Japanese, U.S., and New Zealand road funds were set up, this financing arrangement was referred to as the "user pay" principle. It consisted of two elements: the user paid and proceeds were credited to a special account, or road fund, that was managed separately from the government's budget.

Part III.
An Agenda for Reform

5. Assigning Management Responsibility

Managers of road networks cannot be held accountable for the condition of roads unless responsibilities for managing different parts of the road network and road traffic are assigned clearly. Managing road traffic involves controlling vehicle weights and dimensions, providing road signs and signals, regulating vehicle safety, regulating motor vehicle emissions, managing on-street parking, and controlling road congestion. This chapter discusses how managerial responsibilities are assigned, how they affect overall management of the road network, and how they affect responsibility for managing road traffic.

Basic Principles

The formal way of assigning responsibility for managing a road—which also establishes ownership of the road—is by *designating* the road (box 5.1). The notice that does so cites the act under which the road is to be designated, the location of the road, the responsible road agency, and the functions to be assigned to that agency. In this way responsibility for certain roads may be assigned to a central government agency, a local government agency, a community group, or a private entity (as with private sector toll roads). Responsibilities are normally assigned on the basis of a road's functional classification. As the functional class changes, it should be reassigned from one road agency to another—usually from a lower- to a higher-level road agency, although downward reclassification also takes place.

One of the perennial problems in developing and transition economies is that the road classification system is often out of date. New roads may not have been designated, and changes in the functional class of existing roads may not have been accompanied by reassign-

---

Box 5.1 Establishing the legal status of roads in anglophone countries

In anglophone countries roads fall into two main legal categories—they are either designated or undesignated (the terminology varies among countries, and other terms like proclaimed/unproclaimed, declared/undeclared, and adopted/unadopted are also used to describe the legal status of a road). When a road is designated, the act of designation is published in the government gazette in a notice that cites the act under which the road was designated, the road's location, the responsible highway authority, and the functions to be delegated to that authority. In the case of trunk roads the act cited is usually the Roads and Road Traffic Act (or the equivalent). Urban roads may be designated under the Urban Transport Act (or the equivalent), while other roads may be designated under a variety of other acts, including the Local Government Act, the National Parks Act, the Game Parks Act, or the Private Streetworks Act.

Once a road has been designated, the responsible highway authority is expected to physically mark out the road reserve (to define the land-holding of the highway authority) and to take responsibility for the various functions delegated to it. Roads that are undesignated simply belong to the adjoining landowners who are solely responsible for maintaining them. Under certain circumstances, however, the government may channel funds through a designated highway authority to meet part of the costs of maintenance. When a private road is built to a certain specified standard or is improved to that standard, the highway authority will usually designate it and assign it to a legally constituted highway authority.

Note: This box describes the system of establishing legal ownership in anlglophone countries, though procedures in countries with Spanish, French, and other legal systems are similar.
Source: Prepared by Jeremy Lane and Ian Heggie for this study.

---

ment to the appropriate road agency. Updating the road classification system requires an accurate road inventory and identification of the road agency legally responsible for managing each road. If any roads have not been designated, they will have to be assigned to a legally constituted road agency or, in the case of community roads, to an appropriate community group (like a village council). The inventory may also identify the need to reclassify selected roads, based on changes in their functional class, and to reassign management of some roads from one road agency to another.

The road inventory will normally be used to divide the network into three or four functional hierarchies. The roads can then be grouped into consistent classes for setting common management objectives, construction and maintenance standards, and intervention levels. Countries with relatively low volumes of traffic often group their roads into three functional hierarchies: arterial roads, collector roads, and access roads (TRRL Overseas Unit 1988), while countries with high volumes of traffic usually group roads into four main functional hierarchies with several subdivisions: expressways, strategic routes, distributor roads (including main and secondary distributors), and local roads (including local roads and local access roads) (Local Authority Associations 1989).

The process of assigning managerial responsibility attempts to reconcile three conflicting objectives. First, to the extent possible, it attempts to keep the various functional hierarchies together. Second, it attempts to assign managerial responsibility in a way that is consistent with the country's administrative structure. Since most countries are attempting to decentralize administrative responsibility to reduce the fiscal burden on the central government and strengthen local accountability, this means assigning managerial responsibility for collector roads and local access roads to provincial and district-level governments. Third, it attempts to assign responsibility to agencies that have the financial and technical capacity to manage the roads effectively.

Similar principles apply when assigning responsibility for managing road traffic. Issues that are primarily local in nature (like managing urban road congestion) are normally assigned to local governments, while those dealing with use of the network as a whole (such as regulating vehicle weights and dimensions) are normally handled by the central government. The agency with prime responsibility for managing road

traffic may delegate some of these responsibilities to other levels of government (for example, responsibility for road signs and signals may be delegated to local governments) or to the private sector (for example, responsibility for axle-weight enforcement and vehicle inspection may be contracted out to the private sector).

Managing the Road Network

For the purpose of assigning managerial responsibility, the road network is divided into several administrative classes. Responsibility for managing the roads within each class is then assigned to a public or private sector agency, which becomes the custodian or temporary "owner" of these roads. The way in which these responsibilities are assigned generally depends on the size of the country, the extent of motorization, and the administrative structure of the central and local government.

The simplest management structures tend to be associated with centralized government systems in countries that are relatively small. They have a single-tier structure in which one or more central government road agencies take responsibility for managing most or all of the road network (as in Bangladesh, Jamaica, and Ghana). On the other hand, when the political and administrative system clearly distinguishes between the central and local government, countries tend to adopt a two-tier management structure (as in Latvia, the United Kingdom, and Zambia). Under this system local governments also become involved in management. This model suits countries with extensive road networks that cannot easily be managed by centralized road agencies.

Finally, countries with a federal administrative system tend to adopt a three-tier management structure in which central, provincial (state), and local governments all play a role in managing the road network. The three-tier structure is, however, not universal in federal countries. In some, such as Australia, Canada, and the United States, the federal government delegates the management of most roads to provincial or state governments, and the management structure ends up looking like that in two-tier countries (see box 5.2).

The remainder of this section divides the road network into four administrative classes:

Box 5.2 Jurisdictional control of roads in the United States

Jurisdiction	Rural mileage	Percent	Urban mileage	Percent	Total mileage	Percent
State	692,414	22.3	107,058	13.3	799,472	20.5
Local	2,229,668	71.9	694,728	86.5	2,924,396	74.9
Federal	179,561	5.8	1,292	0.2	180,853	4.6
Total	3,101,643	100.0	803,078	100.0	3,904,721	100.0

Most (74.9 percent) roads in the United States fall under the jurisdiction of local governments (town, city, county). Only 4.6 percent are under the jurisdiction of the federal government. These include roads in national forests and parks and roads on other federal lands and Native American reservations. The rest of the roadways (representing 20.5 percent of total national mileage and including the entire interstate system) are controlled and maintained by state governments.

- National roads, that is, major trunk roads, including expressways and toll roads.
- Regional and rural roads.
- Urban roads, which may also include some toll roads.
- Community roads, tracks, and trails.

The Trunk Road Network

Trunk roads, which typically account for between 10 and 20 percent of the overall public road network, have traditionally been managed through a central government department that forms part of the ministry of works or, in some countries, the ministry of construction. These roads tend to be the most heavily used and the most costly to build and maintain. Although the primary task of this department is to manage the trunk road network, it is often expected to assist local government road agencies and, if needed, temporarily take over management of local government roads. Growing volumes of traffic have also meant that the main road agencies are increasingly involved in constructing and operating high-grade expressways and public toll roads and overseeing toll roads built and operated by the private sector (table 5.1).

Although the functions of the main road agencies have evolved, these changes have not always been accompanied by institutional reforms to ensure that the agencies have a clear idea of their diverse and growing responsibilities and are restructured so that they can discharge these responsibilities effectively. To address this issue, many countries have decided to make the main road agency responsible primarily for managing high-volume (national) roads and expressways. In parallel, a growing number of countries are asking which roads can be realistically managed by the private sector and whether the remaining public sector roads can be managed in a more commercial manner.

Table 5.1 The importance of expressways and toll roads in overall road management
(kilometers)

	Main road agency		Toll roads			Total
	Expressways[a]	Other roads	Publicly managed	Privately managed	Privately owned[b]	public roads
Argentina	600	28,600	0	9,800	0	216,000
France	8,581	29,050	5,562	0	743	966,000
Hungary	378	29,300	0	0	57	158,600
Indonesia	0	17,700	280	0	150	260,000
Italy	894	44,206	0	5,550	0	314,360
Japan	5,860	57,500	8,723	0	0	1,144,360
Korea, Rep. of	0	12,052	1,840	0	40	77,000
Malaysia	575	15,400	0	0	1,010	94,000
Mexico	0	42,928	2,507	3,176	0	303,262
South Africa	615	4,693	0	672	153	525,000
Spain	4,939	22,536	0	0	2,023	343,200

a. Expressways without tolls.
b. Includes joint ventures between the public and private sectors.

This rethinking has led to the establishment of a growing number of autonomous or semi-autonomous road agencies that manage the trunk road network or toll roads along commercial lines (table 5.2). The agencies managing the trunk road network tend to be managed by a public-private board, which includes representatives of road users and the business community; have a chief executive and line managers; operate under a performance contract with the parent ministry; and employ workers under terms and conditions similar to those used in the private sector. The annual performance contract is a key element of the arrangement. It is usually based on a multiyear business plan and spells out the road agency's obligations, strategies for achieving them, performance targets, and procedures for monitoring and evaluation. Such arrangements · assign managerial responsibility in a clear and transparent manner.

Responsibility for managing toll roads is less well-defined. Some toll roads are managed directly by the main road agency, as in Ghana where they are managed by the Ghana Highway Authority; managed by the private sector under a management contract with the main road agency (in Argentina and Malaysia the private sector rehabilitates, operates, and maintains some roads, while in South Africa, they mainly build, operate, and maintain roads); managed through an autonomous road toll agency as in Japan and Korea; or are owned and operated by the private sector under various types of concession agreements (box 5.3).

Between 5 and 20 percent of the trunk roads managed by the main road agency can be operated as toll roads, depending on traffic densities and toll rates (see table 5.1). The revenues from these tolls generally cover the entire cost of operation and maintenance. Usually, less than half of the toll roads cover their capital costs. Those that do not cover all costs are often financed using the loan supportable by revenue approach (box 5.4). The general rule is that a tolled full-standard motorway link will cover all of its costs only when: traffic volumes are at least 10,000 to 15,000 vpd and growing, average toll rates for private vehicles are $0.03 to $0.06 per km, and the concession period is between 20 and 30 years (International Road Federation 1996). As a result, there are relatively few privately owned toll roads. Indeed, it is difficult to let concessions to construct, operate, and transfer more than about 5 percent of the roads managed by the main road agency—and this amount usually accounts for less than 1 percent of the overall public road network. Costs of rehabilitating existing roads can still be economic with lower traffic volumes of around 5,000 to 6,000 vpd. That is why design, build, finance, and

Table 5.2 Countries with autonomous or semi-autonomous main road and toll road agencies, 1998

Established	*Being established*	*Under consideration*	*For toll roads only*[a]
Australia[b]	Lesotho	Kenya	China
Colombia	Malawi	Lebanon	France
Finland	Mozambique	Peru	Indonesia
Georgia	Namibia	Philippines	Italy
Ghana	Nigeria	Romania	Japan
India	Zambia	Tanzania	Korea, Rep. of
Ireland		Uganda	Malaysia
Latvia		Zimbabwe	Spain
New Zealand			Thailand
Sierra Leone			
South Africa[c]			
Spain[d]			
Sweden			
United Kingdom[e]			
Yemen			

Note: Pakistan, Russia, and Sri Lanka (not listed) have road authorities in name only.
a. Both public and private toll road agencies.
b. Some states have established semi-autonomous highway authorities.
c. To be established as of end-March 1998.
d. Some regions only (for example, Andalucia).
e. Highways Agency in England.

Box 5.3 How different countries manage toll roads

KOREA (DESIGN, BUILD, FINANCE, AND OPERATE): Korea Highway Corporation is a public corporation charged with constructing, reconstructing, and maintaining national expressways, all of which are operated as toll roads. By the end of 1996 Korea had 20 expressways with a total length of 1,840 km. There is a bold plan to add a further 1,800 km and to expand 500 km of existing roads by 2004. Then, expressways will account for about 25 percent of the national road network and nearly 5 percent of the overall public road network.

Average traffic volume on expressways is more than 44,000 vehicles per day. Revenues are pooled, and there is a unified toll fee of $0.031 per km for cars, $0.035 per km for buses, and $0.065 per km for trucks. Tolls were originally introduced to generate sufficient revenues to cover the costs of operating and maintaining the express-way network. But shortage of public revenues has meant that Korea Highway Corporation is increasingly required to contribute part of the costs of new expressway schemes. The government provides the balance of the capital in the form of equity.

Korea Highway Corporation operates on a break-even basis, and tolls are adjusted from time to time to ensure that revenues cover costs, including redemption of loans and payment of interest on government loans and corporate loans and bonds. Efforts are currently underway to persuade the private sector to build and operate selected toll roads under concession agreements. The first such project is being implemted, and it is expected that several more will be agreed during the next few years.

SOUTH AFRICA (PRIVATE FINANCE AND OPERATE): South Africa currently has 10 continuous toll roads totaling 709 km in length. Average daily traffic taken over all toll plazas was 8,892 vpd during 1995–96, with variations from a low of 2,292 vpd to a high of 26,143. These limited-access freeways are either new or have been significantly rebuilt. Almost all work, including construction, maintenance, and operation of toll plazas is carried out by the private sector under open-tender contracts. The roads are overseen by the Department of Transport, which acts as the administrative arm of the South African Roads Board. The legislation under which these roads are tolled stipulates, among other things, that the Ministry of Transport determines toll tariffs based on recommendations submitted by the South African Roads Board.

The toll tariffs are raised regularly to account for inflation. The toll system operates on an open basis (that is, motorists can use sections of road between toll plazas without paying the toll). Regular users are given substantial discounts through concessions to local residents and frequent-user cards. In addition to the above roads, there is one private sector toll road carrying a government guarantee and another under consideration without any guarantees.

ARGENTINA (PRIVATE DESIGN, BUILD, FINANCE, AND OPERATE): The Dirección Nacional de Vialidad is responsible for 38,000 km of roads. Many of these needed to be rehabilitated, and the Dirección Nacional de Vialidad decided to invite bids from the private sector to rehabilitate, operate, and maintain up to 10,000 km of this network and to recover the costs through tolls. The roads were divided into 20 corridors varying in length from 300 to 1,000 km. Improvement needs were identified for each corridor, and a schedule was established for the work to be carried out by the concessionaire.

Some work had to be done before any tolls could be collected, while other work had to be completed within 36 months. After a bidding process, in which concessions were awarded to bidders offering the highest lump-sum payment for the concession, 14 concessions were awarded to 13 consortia covering 9,800 km. The government remained owner of the roads and set the basic tariff (indexed on the basis of the cost of living and the exchange rate). The concessions were for 12 years, and the roads had to be maintained to specified standards. The initial tariff was set at an average rate of $0.015 per km.

The concessions operated for only about six months—the tariff rose rapidly, a public outcry arose, and the concessions were suspended. After further negotiation, the concessions were reinstated based on an average toll rate of $0.01 per km, the lump-sum payments were abolished, and the government agreed to pay an annual subsidy of $57 million to the concessionaires.

---

operate contracts involving partial finance by the private sector are becoming so popular.

Regional and Rural Roads

Regional road networks are often large, and it is therefore feasible to manage them through a road agency analo-gous to the one managing the main trunk road network. But the same is rarely true of the rural road network since rural municipalities and rural districts are often small and lack the capacity and resources to manage the local road network effectively. The difference is essentially a matter of scale. The local road networks are too small to justify

---

Box 5.4 Financing toll roads in South Africa: the loan supportable by revenue approach

In South Africa none of the initial toll road schemes implemented by the Department of Transport on behalf of the South African Roads Board were wholly self-financing. The concept applied in designing the initial schemes was called the loan supportable by revenue approach. The loan supportable by revenue is determined by calculating the present value of the project over 30 years, using a projected real interest rate, expected traffic growth, and forecast expenditures. This figure determines the size of the loan that can be repaid from toll revenues over 30 years at a borrowing rate above the rate of inflation. The balance of the capital is provided in the form of a long-term loan from the National Road Fund. These loans (together with any interest determined by the Board) are repayable only when all private sector loan obligations have been met.

Experience to date indicates that traffic growth has exceeded initial expectations, and real interest rates have been higher than forecast after an initial period of being less than forecast. The loan supportable by revenue calculations are revised on a regular basis to ensure that all private sector financial obligations will be met. Although the toll roads incurred deficits during their early years, a combination of inflation-linked tolls and traffic growth is expected to enable the toll roads to break even within 20 years and fully repay private sector loans within 30 years.

Private sector financing involves both capital and money market loans using a variety of instruments. The total borrowings as of March 31, 1997 amounted to $645 million in the capital market at nominal values with redemption dates up to and including 2015, and $99 million in the money markets. Overall financing costs in 1996–97 were 13.5 percent. No interest was paid on National Road Fund loans, amounting to more than $390 million.

Source: Prepared by Albert du Plooy for this study.

---

recruiting the skilled personnel needed to plan and manage a road to a technically acceptable standard. This problem raises serious questions about how to assign legal responsibility for managing such networks.

Countries have typically attempted to deal with the above problem by:

• Reclaiming legal responsibility for rural roads and transferring it either to a central government ministry or to a special-purpose rural roads agency.

• Persuading (or requiring) local governments to hand over responsibility for implementing their road programs to a specialized project implementation agency.

• Persuading local governments to join together to form larger operating units for the purpose of managing their road networks, that is, to form joint-services committees.

• Persuading local governments to contract out the planning and management functions to consultants.

Central government management. The first model, which is used in both Bangladesh and Ghana, solves the problem of scale and also provides access to central government budgetary resources, at least for capital works (box 5.5). The Department of Feeder Roads in Ghana has often been held up as a model of best practice for managing feeder roads through a central government department. It was established in 1981 when it became clear that the Ghana Highway Authority could not manage the rural road network. The Department of Feeder Roads overseas a network of 21,830 km, is highly decentralized, and does most work under contract—often using labor-based methods. But now it will have to transfer many of its responsibilities back to the rural district councils as part of the government's overall decentralization program. It is still unclear how this will be done and whether the Department will remain a contractor managing the local networks on behalf of the districts or will be restructured into a small central government department that monitors local road works, provides advice, and ensures quality control.

A central government department can successfully manage rural roads, but such an agreement generally represents only an interim solution to the problems of weak local government capacity. Centralization, particularly in large countries, bypasses the local government road agencies, further weakening them. Meanwhile, remote central government agencies rarely consult local communities about priorities to the extent that they should and are less concerned about long-term sustainability. But these problems can be avoided in smaller countries, given proper legislation requiring formal consultations and local endorsement of road plans.

Project implementation agency. The second model is used extensively in francophone Africa. Such an implementing agency is generally referred to as AGETIP (*Agence d'execution des travaux d'interet public contre le sous-emploi*), and they have been set up primarily to execute donor-financed infrastructure projects. The agency generally has a board composed of well-known figures (none of whom are government representatives), a general manager appointed by the board, other line managers (administrative, financial, and technical managers), and staff who are paid market-based salaries. The agency is set up as a private, nonprofit association and pays no taxes. It works on behalf of local authorities who delegate certain functions. The local government usually reserves the right to select projects. The project implementing agency then recruits consultants to carry out detailed engineering, invites bids and awards contracts for both supervision and implementation of works, manages the contracts, and pays the contractors directly from a special account opened in its own name. The agency is subject to bimonthly management and financial audits, and an annual technical audit. The overhead cost of the AGETIP in Senegal (excluding the fees paid to consultants for designing and supervising works) is about 5 percent on a turnover of $55 million (330 projects).

The advantages of an AGETIP are that it avoids cumbersome government procurement regulations, stream-lines payment procedures, and pays high salaries, therefore attracting well-motivated, high-quality staff. Its "corruption-free" procedures have also allowed them to complete most projects on schedule with a cost-overrun of just more than 1 percent of the portfolio (in contrast, cost overruns in public procurement in the same countries average 15 percent). The AGETIP routinely obtains unit prices 5 to 40 percent lower than those obtained by the government through official bidding.

The disadvantages are that the contract between the AGETIP and the government is not itself subject to competitive bidding, the AGETIP is almost entirely dependent on continued donor funding, and it probably hampers development of the local consulting industry by creaming off staff and monopolizing all contract management work under its tax-free operating environment. The AGETIP can nevertheless play an important role, particularly as an interim solution in economies where the local consulting industry is relatively undeveloped. Also, in the longer term it could evolve into a contractual arrangement awarded on the basis of competitive bidding.

Joint services committees. The third model, in which several local government agencies cooperate to procure goods and services on behalf of all their members, is fairly common in industrial countries, as well as in Jordan

---

Box 5.5 The Local Government Engineering Department in Bangladesh

The Local Government Engineering Department started in the 1970s as a small cell in the Local Government Division. Today, the Department is a dynamic agency with about 700 engineering staff who are led by an energetic director. It is the second largest of the departments implementing the country's Annual Development Program. It is responsible for construction and maintenance of rural infrastructure, including about 95,000 km of rural roads, most of which are earth. Almost all works are contracted out to the private sector, with smaller tasks going to worker groups. The Department is praised by both service users and donors as one of the most efficient and effective government organizations in Bangladesh. The institutional aspects contributing to its successful operation include:

• Decentralization—90 percent of the staff are located at the *thana* (subdistrict) level.

• Professionalism—the Department is well-known for its highly qualified professionals and its emphasis on continual skill upgrading.

• Monitoring system—a computerized management information system has been in place for more than a decade.

• Informal decisionmaking—the Department has bypassed the time-consuming practice of processing decisions through many layers of the bureaucracy.

• Leadership—strong, consistent leadership has provided continuity and motivated the staff.

• Teamwork—the staff have clearly defined work objectives and a keen sense of achievement.

• Sense of mission—the Department has vigorously pursued its mission of "serving the people at the grass-roots."

Source: World Bank (1996a).

and South Africa. The arrangements are normally referred to as joint-services committees. Joint-services committees are used mainly to acquire scale. By pooling their resources, individual agencies are better able to plan and manage their affairs and let larger and more cost-effective contracts for procuring goods and services.

The group of local government agencies generally assigns the tasks of organizing procurement and supervising implementation to one of their members, to a higher level of government, or to a local consultant (see the fourth option below). The collaborative arrangement is sometimes made on an informal basis, although there usually must be a written agreement among the participating parties when the joint services committees becomes involved in activities like road maintenance, particularly when cost-sharing with the central government or a road fund is involved. These legal agreements tend to be quite complex, especially when the local government agencies are controlled by different political parties. However, they do build local capacity and identify who is responsible for what.

Private consultants. The fourth model, which involves contracting out planning and management functions to consultants, is growing in popularity. Under this arrangement the local government agency remains responsible for managing the road network— and often does so through a small in-house client unit—while the actual planning and management are delegated on a competitive basis to a technically qualified third party. In industrial countries this model is being adapted as part of the process of redefining the role of government and to increase efficiency. In developing and transition economies its main attraction is that it enables small road agencies to gain access to people with high-quality technical planning and management skills without taking the decisionmaking power away from the local government. This model is being used in some small municipalities in the United States, at the county council and district levels in the United Kingdom (box 5.6), for some rural district councils in Tanzania, and for all district councils in Zambia (box 5.7).

Urban Roads

Urban roads can be grouped into three broad classes: roads in large metropolitan areas (generally comprising several large and medium-size cities), roads in large cities, and roads in small towns. Each class faces unique problems, and they thus tend to be managed differently. Some of these differences are logical and easily understood, while others have been arrived at

---

Box 5.6 Contracting out planning and management of local government roads in the United Kingdom

Prior to 1980 local governments in the United Kingdom typically employed their own workers to advise, manage, design, and deliver projects and services to their customers. This system changed in 1980 when new legislation compelled councils to let large construction contracts under competitive tenders and required in-house direct labor to operate as quasi-companies and produce a small profit at the end of each year. The regulations were subsequently tightened to require that all new works undertaken by direct labor be subject to competitive tendering.

Some councils later decided to seek competitive tenders for provision of local authority services (such as garbage collection). The central government followed up on this by introducing competitive tendering for all such services. By the early 1990s the concept of competitive tendering was well-established, and the government decided that competitive tendering should also apply to professional services (including planning and engineering services). The law required that by April 1, 1996 at least 65 percent of all construction-related services be subject to the market test, that is, compulsory competitive tendering.

Under this arrangement the local authority's in-house organization is split into two parts. One part remains in-house as a "client unit," while the other becomes a "contractor," consisting either of previous in-house staff who have to bid competitively for the work or of an outside bidder who wins the contract. By early 1996 this process was well under way, and many councils had decided to contract out most construction-related services rather than allow in-house staff to bid for the work (the process of contracting out was referred to as *externalization*). Berkshire County Council externalized highways and planning, Westminster City Council externalized planning and transportation projects, the London Borough of Ealing externalized all technical services, and other local authorities externalized selected functions like engineering, design and materials, highways, and traffic management.

Box 5.7 Contracting out planning and management of rural district roads in Zambia

Zambia has 15,980 km of rural roads managed by 48 rural district councils. When the National Roads Board took over management of the road fund, they invited the districts to submit proposals for maintenance programs. The programs submitted left much to be desired. They typically consisted of a list of road names coupled with a financial figure. There was no assessment of road conditions, no details of the proposed maintenance works, no indication of how the work was to be done, no specifications, and no contract documents.

The National Roads Board therefore approached the Association of Consulting Engineers of Zambia and asked if their members would help to prepare acceptable maintenance programs on behalf of the districts. The Association agreed to this on the basis of a terms of reference that required the consultants to work with the councils and:
• Agree on a road maintenance program and order of priority.
• Agree on procedures to be followed in calling for tenders.
• Assist in tendering and evaluation.
• Negotiate with winning tenderers.
• Agree on the selection of contractors.
• Negotiate and agree on the terms of contract.
• Ensure that implementation adheres to a set time frame.
• Assist and monitor progress of road works to ensure total quality management.
• Certify payments at each stage of road works if the quality of work is satisfactory.

• Help councils to control and reduce costs of road works and maintenance through consulting services.
• Advise councils on undertaking preventive maintenance activities to promote the quality and life span of road infrastructure.

The Association of Consulting Engineers also agreed to introduce performance indicators for their members, including quality of the program drawn up, time frame for implementation, unit costs of road works undertaken, quality of work done, volume of work done, public relations and involvement of the local community, and preventive maintenance activities. By 1997 the performance indicators had already resulted in the termination of at least one contract with a consultant.

Bids were invited from members of the Association of Consulting Engineers to act on behalf of all the districts in each province. The National Roads Board and the concerned districts selected firms, and one firm was appointed per province under a local government service contract. Within each province a single consulting firm now works with the districts, prepares their maintenance programs, prepares bid documents, helps them select contractors, and supervises implementation. The National Roads Board fully reimburses work done under contract. When a district does work using in-house staff and equipment (mainly pothole patching), the National Roads Board pays for materials only after certification that the work has been done according to specification.

---

through a process of incremental change. The engineering criteria used in interurban transport may become less important than other criteria, such as pedestrian needs, traffic congestion, pollution, and the expectations and rights of city dwellers. These considerations can modify decisions that would otherwise be made on purely engineering grounds.

Roads in large metropolitan areas. The simplest model for large metropolitan areas is an area-wide authority with full responsibility for roads and highways. This model has the advantage of economies of scale and mechanisms to direct funds to the areas of most need. But not all parts of the metropolitan area are likely to have or perceive that they have the same problems. Divisive arguments between different parts of the metropolitan area may arise, along with accusations that decisions are being made at the wrong level.

In a second model the constituent parts of the metropolitan area have full authority for their own network of roads and highways. These parts are normally large enough to achieve the necessary economies of scale and, in addition, can deliver services closer to the population. Unfortunately, conflicts can arise when decisions are made on cross-boundary or strategic routes. Such decisions can have far-reaching consequences in neighboring areas—which have no formal say in the decisionmaking process.

Recognition of this problem has led to the adoption of a third model in which a strategic authority takes responsibility for certain strategic roads or functions. This authority can be a metropolitan-wide body, such as an elected authority, or a regional or state body empowered to fulfill this role. In a technical sense this system has much to recommend it. Unfortunately, conflicts between local bodies and the strategic body then arise because the strategic body has a bias to manage, operate, and finance the road in the interests of strategic traffic. The local authority may naturally wish to restrict traffic on such roads or to allow more frontage development to the detriment of the free flow of traffic.

In a fourth model the strategic authority comprises representatives from each of the local authorities, creating a mechanism for dealing with differences. Still, local agendas have historically predominated in such joint bodies and the only way to progress is through a process of deal-making.

The models mentioned above delineate ultimate responsibility for highway functions. The actual delivery of services can also involve different mechanisms. The most straightforward is the one in which the road agency is responsible for delivering the entire service, either through its directly employed work force or through one or more contractors. This method allows little ambiguity concerning responsibility and is easily understood by practitioners and the public. A less straightforward mechanism is one in which one authority, normally the strategic authority, delivers the service using another authority as its agent (box 5.8). The rationale behind this method is that the local authority will be able to offer economies of scale by incorporating the services required for the relatively short lengths of strategic roads into whatever arrangements it has for delivering services on the much longer length of its own roads.

While this method does have an advantage in obtaining competitive prices, it suffers in other respects. In particular, the local authority tends to act as though the strategic roads were its own, which can create conflicts between the local authority and strategic authority. There is also confusion over who has responsibility for what, especially in the eyes of the public, who invariably find this arrangement difficult to understand.

Roads in large cities. The options outlined above for metropolitan areas may also apply to large free-standing towns and cities. But, in reality, the arrangement is normally one in which the urban area has total responsibility for its roads or is in some way linked to the local region or state, which plays a strategic role. This linkage can take several forms (as outlined for metropolitan areas). Further, sometimes under agency arrangements the responsibility for roads falls to the urban

> **Box 5.8 The agency system in the United Kingdom**
>
> In the United Kingdom an agency system operates in which the highway authority, normally the county council, devolves the delivery of certain functions to a lower tier of local government—in this case a district. The exact arrangements vary among and within counties. In general, such arrangements work best in urban areas, and the bigger is the urban area, the bigger is the number of functions that can be devolved. How agencies should operate was the subject of a government report, and the following guidance was given on functions that might be suitable for an agency. Maintenance, grass cutting, sweeping, gully emptying, relaying footway, white lining, fencing, and cleaning signs, bollards, and traffic signals may be carried out by an agency whose district is highly or sparsely populated. Patching, potholing, curbing, resurfacing, and reconstructing should only be undertaken by the agency if the district has a population of more than 100,000. Similarly, street lighting is appropriate only for a district of more than 180,000. Winter maintenance is not dependent on district size, provided the routes fall largely within the district. Lastly, drainage, bridges, earthworks, and electronic maintenance of traffic signals should be carried out by the county.

area, but the urban area must conform at least in part to policy guidelines set down by the region or state. This set up leaves the urban area free to provide value for money by planning the delivery of services however they see fit (box 5.9).

Roads in small towns. In small urban areas the responsibility for managing roads can rest with a nonurban authority, such as the region or state, the rationale being that the urban area is too small to carry out or procure the necessary expertise. Alternatively, many countries operate a system in which the urban authority has responsibility for all roads, or all nonstrategic roads, that pass through their area. In another system certain functions are the responsibility of the urban area and some the responsibility of the region. For example, routine maintenance activities like sweeping and pothole repairs may be the responsibility of the urban area, while larger-scale works are the responsibility of the regional body.

Service delivery in small urban areas takes many forms. It is typical to find that only certain core func-

Box 5.9 Funding relationships in urban areas

Box 5.9 Funding relationships in urban areas

In several African countries the funding relationship between the city and the central government is through a bidding process. In general, over an annual cycle the city determines which roads need repair and presents this bid to the central government. The central government considers bids from all over the country and decides which resources should be allocated to which cities. This practice varies according to the level of detail. In some cases individual, relatively small schemes are individually approved and funded. In other cases only large schemes are individually approved and funded, while smaller schemes and general works are jointly allocated funds.

Such generic systems are common in Europe. These mechanisms can be just as complex as the managerial relationships. It is not uncommon to find that funding is made available from all levels of government and from the private sector in one form or another. The availability of such funds, the ease with which they can be obtained, and the conditions attached can distort spending decisions. In addition, the public has difficulty understanding such systems and hence bringing pressure to bear to change policy decisions and introduce efficiencies.

These funding mechanisms can also distort managerial responsibilities. It is easy to blame inadequate resources from a different level of government as the cause of poor maintenance. But when funding and responsibility come from the same body, there is a clear imperative to maximize efficiency in all aspects of service delivery. There is no one else to blame.

Source: Prepared by the Babtie Group for this study.

tions are provided in-house because of the size of the operations, or core functions are provided under long-term contracts (as in the case of many small Finnish communities that contract out all road works). Other less-frequent operations may be carried out through individual contracts, framework agreements, or call-off contracts.[1] This system allows access to organizations with the necessary expertise without having to maintain in-house resources all year.

Some small urban areas form consortia with other adjacent bodies to provide services on a lead-authority basis—one town may deliver a particular service to a group of like-minded towns. Other towns may take on similar roles for other services. This system can work effectively if authorities are like-minded, but is difficult to orchestrate if authorities have markedly disparate

aims and objectives. Taking this process one step further, authorities could enter into quasi-legal arrangements, often by letting contracts for the delivery of a specified service to one another or to an outside body. This arrangement is different from the joint bodies formed by parts of metropolitan areas. In that case the joint body was responsible for policy, while in this instance it is responsible for implementation. It should also be noted that toll roads—with either real tolls or shadow tolls—can be superimposed on any of the above arrangements by a strategic, regional, state, or local authority.

Finally, all urban areas normally have some undesignated or private roads that are normally the responsibility of the adjoining landowners. In many cases such roads form the majority in urban areas even though they have a tenuous legal status. Often the adjoining landowners have rights, privileges, and theoretical responsibilities that are honored more in their breech than in their observance. The urban authority has certain step-in rights, but does not often exercise these rights—deliberately, because of a shortage of funds, or because of the sheer complexity of the legal and bureaucratic process involved. Unfortunately, an impasse can arise and last for many years.

The Lowest Level of the Road Network
Managerial responsibility is not well-defined at the lowest level of the road network. Some roads may be nominally placed under the jurisdiction of village councils (or the equivalent), while others may be treated as private roads and left in the hands of adjoining landowners or state enterprises (as with agricultural roads in Russia before the restructuring of collective farms). At this low level there are few technical skills and an acute shortage of funds for maintenance, upgrading, and new works.

Some countries attempted to address this issue by creating a special class of roads—by amending the basic road legislation or by passing a Private Streetworks or Road Cooperative Act—to which public funds can be channeled on a cost-share basis. The key to these arrangements is setting up an institutional arrangement that offers appropriate incentives and clearly defines who is responsible for what. This so-called "private road" option is particularly appealing as devolution of government becomes global.

The private road arrangements in Canada, Finland, Norway, Sweden, and the United States have common elements. Their main objective is to persuade local people—individuals, villages, or groups of villages—to accept responsibility for managing their own roads. Persuasion usually requires that there be an incentive system, access to advice and technical assistance on road management, and technical and financial oversight to ensure accountability of public funds. Incentives normally take the form of cost-sharing arrangements. Higher-level government pays part of the construction and maintenance costs, and the adoptive owners of the road pay the balance. But, to participate in the cost-sharing agreement, the individual or group must formally apply to join the agreement and abide by its rules. In Finland they do so by forming a road cooperative (it is compulsory for property owners to join an established cooperative if they maintain a road that provides the only access to their property). In Ontario they form a local road board, while in Lesotho they form a village development committee.

There is obvious need for technical advice at the local level, especially in developing and transition economies. Local villagers need to be trained in planning, carrying out road works, and dealing with unexpected problems during implementation. Such instruction is usually provided through a local planning agency and through the agency responsible for managing the main or rural roads. Finally, there is the question of oversight, the need to ensure that road works are carried out to agreed standards. There is also a need to ensure that government funds are properly accounted for. Technical supervision should ideally be provided by the agency responsible for rural roads, though there is generally no need for financial oversight when the local contribution takes the form of volunteer labor. When cash is involved, there must be an oversight arrangement like that in Finland. The Finnish system is well-developed and might be used as a model for developing and transition economies (box 5.10).

Box 5.10 Managing private cooperative roads in Finland

Finland has 78,000 km of public roads, 24,000 km of city streets, and about 280,000 km of private roads that are maintained by adjoining landowners or people living alongside the road. Private roads with more than one owner can be managed as road cooperatives. By 1997, 104,000 km of all private roads had been legally constituted under the Private Roads Act as cooperative roads. These roads carry an average of 45 vehicles per day, and 99 percent have gravel and earth surfaces.

The Private Roads Act requires that the cooperative stipulate right-of-way, ownership, and the formula for distributing maintenance costs among both road users and the adjoining property owners. The cooperative is responsible for arranging maintenance and may either pay its own members to do the work or use a contractor. Each cooperative must hold an annual meeting and must elect a chairperson, a secretary, a trustee, and other office-holders to manage their maintenance operations. The trustees charge about $200 annually to manage an average-size cooperative. The cooperative sets its own maintenance fees, accepts new members, and is responsible for having the previous year's accounts audited. Membership is compulsory for property owners who use the road. Maintenance costs are shared among members, depending on the size of their property and the amount of traffic they generate.

The government supports maintenance of cooperative roads provided that a formal cooperative has been established, the road length to a permanent residence is at least 1 km, and there are at least three estates with permanent residents alongside the road. Each municipality has its own rules for supporting cooperative roads under their jurisdiction. In 1997, 87,000 km of the 104,000 km legally designated as cooperative roads received public support from the government, a municipality, or both. The support was given to 17,400 cooperatives with 392,000 members. In 1990 the government provided about $30 million to support cooperative roads, municipalities provided $40 million, and the remaining $50 million came from members of the cooperatives.

Government support is channeled through the Finnish Road Administration (FinnRA) and is allocated to each qualifying road on the basis of traffic volume and number of permanent households served. The amount of government support is adjusted for climate and average income. Additional support may be granted to cover exceptional items. A FinnRA supervisor inspects the qualifying roads once every two to three years and transfers road maintenance know-how through an annual meeting with the road cooperative. FinnRA's administrative costs of managing cooperative roads is about $60 per road cooperative per year.

Source: Isotalo (1995).

Transferring managerial responsibility to local communities through road cooperative arrangements ensures that most roads receive regular maintenance; may reduce the number of roads for which local governments are responsible, making it easier for them to manage the other roads under their jurisdiction; ensures that priorities reflect local expectations; and can be used to stimulate local employment and to improve the efficiency of local road works. For example, road cooperatives in Sweden have managed to reduce their maintenance costs to about half that spent by the Swedish National Road Administration on similar public roads. This savings is partly due to lower (more appropriate) maintenance standards, but also comes from lower overheads and use of otherwise idle farm equipment. Some of the road cooperatives have become so good at carrying out maintenance that the Swedish National Road Administration uses them to maintain some of the low-volume roads on the public road network.

The above arrangements offer a number of advantages. First, they assign managerial responsibility to a clearly identifiable entity. Second, when the arrangements are backed by legislation, they also assign legal responsibility for managing the roads. Third, they introduce mechanisms for providing technical advice and oversight to ensure that government funds are properly accounted for. Finally, they ensure that all parts of the road network, not only those owned by the government, are regularly maintained.

Managing Road Traffic

The concern here is with responsibility for those aspects of vehicle traffic that affect management and financing of roads. Assigning these responsibilities is fairly obvious since they are widely recognized as an intimate part of overall road management. Road signs and signals fall into this category (they are usually specified under international agreements), as do road design standards. These are central government responsibilities and will usually be delegated to the agency responsible for managing the main road network or the organization managing the road fund. The agency with prime responsibility may nevertheless delegate some of these responsibilities (road signing in urban areas, for example) to lower-level road agencies.

At the other extreme it is clear that control of parking, particularly on-street parking, and handling of urban road congestion should be assigned to the agencies responsible for managing the urban road networks. These two areas are closely linked, tied to the issue of urban road safety (particularly pedestrian-vehicle conflicts), and are essentially local in nature. They are therefore generally dealt with most effectively by local governments in conjunction with local law enforcement officials. Traffic regulations, particularly those dealing with routing heavy vehicles in cities, are also largely a local matter. They either form part of local traffic management or, in the case of heavy vehicles, are intimately related to vehicle impacts and construction of local relief roads and urban bypass schemes. Although the broad framework for these regulations should be set by the central government, their detailed application must be delegated to the agencies responsible for managing urban road networks.

More problematic are the questions of vehicle weights and dimensions, vehicle safety, vehicle emissions, and the environmental impacts associated with roads and road traffic. Overweight vehicles damage the road pavement and increase road maintenance costs. What are considered permissible vehicle dimensions affect road design standards and hence construction and maintenance costs. All road agencies have a vested interest in seeing that these regulations are well-designed and consistently enforced. Axle-weight regulations are the most important, although they are difficult to enforce. Weight standards are generally promulgated by the central government transport ministry, while the regulations are administered by a wide array of other agencies: weighbridges are operated by the main road agency, penalties are issued by traffic commissioners, and law enforcement is handled by the police.

Reviews of past enforcement efforts have pointed to three areas of weakness: lack of clearly assigned responsibilities, weak enforcement agencies, and resistance by the road transport industry. But the underlying problem is lack of appropriate incentives (Heggie 1991a). The road transport industry sees no direct connection between the damage done to the road pavement and the road-user taxes and charges they pay to the government, while the road agency is reluctant to spend its own time and money on administering the weighbridges when all the revenue from penalties accrues to the treasury.

Part of the solution lies in clearly assigning responsibilities among the various agencies administering axle-weight regulations. In addition, assignment should be supported by other actions ensuring that road users understand the relationship between overloading and pavement damage, the road transport industry agrees to work with the road agency to fairly enforce axle-weight regulations, and the main road agency has an incentive to enforce axle-weight regulations. Several road agencies have persuaded the road transport industry to work with them in an effort to control overloading (see box 6.1).

Getting incentives right is more difficult. There are encouraging signs from countries with commercially managed road funds: private sector board members are beginning to see the direct link between the damage caused by overloaded vehicles and the road tariff that their members must pay. As a result they are becoming increasingly involved in controlling overloading. For example, the road funds in Yemen and Zambia are now taking over management of weighbridges, and the revenue from fines is going into the road fund.

Regulations governing vehicle safety and vehicle emissions are usually administered in conjunction with the issuing of vehicle licenses. In some countries (Finland and the United Kingdom) this is done by the central government, while in others (South Africa and nearly all federal countries) it is handled by local governments. Administration is generally the responsibility of the transport ministry (licensing branch) or the local tax office. Since between 35 and 50 percent of the vehicles in a typical developing country are unlicensed, uninsured, and uninspected, vehicle safety and emissions regulations, where they exist, are not particularly effective (Heggie 1992). Road agencies nevertheless have a clear interest in vehicle safety since it affects the use of the road network and has an important impact on road safety—for which most road agencies have some responsibility. It may therefore be worthwhile to assign responsibility for regulating vehicle safety to the agencies charged with managing the road network.

And since this involves vehicle inspection, to also assign them the role of administering vehicle emission regulations. However, it is only worthwhile to make these assignments if the agencies have the skills and resources to carry out the inspections properly and can compel compliance.

Finally, there is the important question of assigning responsibility for dealing with the environmental impacts associated with roads and road traffic. New road schemes may inadvertently damage ecologically sensitive areas, destroy property, displace people, or disrupt established settlement patterns in both urban and rural areas. The road agency should, at least in principle, be assigned the responsibility of ensuring that potentially beneficial environmental impacts associated with new road schemes are realized, that adverse environmental impacts are minimized, and that any remaining impacts are considered acceptable by the public. The usual way of doing this is ensuring that major road schemes are subjected to some form of environmental impact assessment involving public consultation. Such procedures are now routinely used throughout Europe and North America, and in countries like Brazil, Chile, China, and South Africa. Still, the consultation process must be strengthened and extended to other developing and transition economies.

The road agency normally bears less responsibility for the environmental impacts associated with road traffic. It generally has little control over vehicle performance, which is mainly affected by the tax structure (which influences the size of vehicles, their age, the type of fuel used), vehicle emission regulations and inspection procedures (which affects their state of repair), and the quality of imported and locally refined fuel. These responsibilities are normally assigned to the ministries of environment, energy, and finance. Still, the road agency can play a secondary role in helping to mitigate the adverse environmental effects associated with road traffic. They can install noise barriers, use porous asphalt, plant trees, and landscape the road right-of-way. A road agency should thus be responsible for carrying out environmental assessments on major road schemes, using them as the basis for undertaking public consultations, and possibly implementing remedial measures to mitigate the adverse environmental impacts associated with road traffic.

Key Conclusions and Recommendations

The above discussion leads to the following suggestions for clarifying and assigning responsibily for managing different parts of the road network:

• The formal way of assigning responsibility for managing a road is by designating the road. Responsibilities are normally assigned on the basis of the road's functional classification. Managerial responsibility may be assigned to a central government agency, a local government agency, a community group, or a private entity.

• Since the road classification system in developing and transition economies is often out of date, the government may need to revise the road inventory. The inventory can then be used to reclassify the road network and identify which road agency has been assigned legal responsibility for managing each of its parts. Roads that have not been designated must be assigned to a legally constituted road agency or to an appropriate community group. Some roads must also be reclassified, and managerial responsibility reassigned from one road agency to another.

• Responsibility for managing the main trunk road network will normally be assigned to a central government road department, except in federal countries, where the provinces or states may be the designated highway authorities.

• The main trunk road network is increasingly being managed by autonomous or semi-autonomous road agencies that operate along commercial lines. Likewise, roads carrying large volumes of traffic are increasingly being managed as toll roads, either by the main road agency, autonomous public toll road agencies, or the private sector under various forms of concession agreements. These developments should be encouraged since they promote managerial responsibility.

• Large regional road networks tend to be managed in the same way as the main trunk road network. Smaller networks, which lack the scale needed to support a technically competent local road agency, tend to be managed through a central government department, a project implementation agency (AGETIP), or a joint-services committee, or by contracting out the planning and management of roads to consultants.

• In the case of rural roads joint-services committees and contracting out the planning and management functions to consultants appear to offer viable long-term solutions to the problems of weak technical capacity. The AGETIP model is an alternative that could evolve into a suitable long-term solution.

• In large metropolitan areas roads may be managed by a metropolitan-wide authority with full responsibility for all roads and highways; by the different political jurisdictions making up the metropolitan area; by a strategic authority—an elected authority or a regional or state body—that takes responsibility for certain strategic roads or functions, but leaves all other roads and highways under local jurisdictions; or by a strategic authority made up of representatives from each of the local jurisdictions. Works may be implemented by each body using its own directly employed work force and/or contractors, or by the strategic authority on an agency basis.

• In large free-standing towns and cities the road network may also be managed along the same lines as in large metropolitan areas. But, typically, the urban area has total responsibility for its own roads, receiving varying strategic inputs from the local, regional, or national governments. The urban area is usually free to deliver services however it sees fit, although they may be subject to guidelines laid down by the regional or national government. The tendency is to use contractors for most of the work.

• In small urban areas, where the urban government often lacks the scale necessary to manage its own road network, responsibility for managing roads may rest with a regional or national agency. Alternatively, the urban authority may be responsible for managing non-strategic roads or for handling a limited range of functions (street cleaning, pothole repairs), while a regional or national agency may be responsible for strategic roads or larger-scale works. Because of the small scale, many core functions will likely be contracted out, though some may be provided in-house.

• At the lowest levels of the road network, where roads are often undesignated, it is best to offer incentives to local residents—individuals, villages, or groups of villages—to accept responsibility for managing their own roads. This includes having them adopt the roads (by forming a local road cooperative or the equivalent), receive grants from the government or the road fund to meet part of the costs of maintenance and improvement, gain access to technical advice, and be subject to some form of oversight to ensure that any government funds are properly accounted for and that technical standards are met.

• The main road agency should be assigned all regulatory responsibilities affecting the entire road network (such as design standards, signs, and signals), even though it may delegate some of these to other road agencies or competent bodies.

• Urban road agencies should be assigned responsibilities (control of parking and congestion and routing of heavy vehicles) that significantly affect urban areas.

• The main road agency should be responsible for enforcing axle-weight regulations, ideally in conjunction with the road transport industry. The main road agency could also carry out vehicle safety and emission inspections, provided they have the skills and resources to do them properly.

• All road agencies should be made responsible for examining the potential environmental impacts of new road schemes and should be required to ensure that adverse effects have been minimized and that remaining impacts are acceptable to the public. Road agencies should also be required to consider remedial measures to mitigate the adverse environmental impacts of road traffic, including provision of noise barriers.

Note

1. Call-off contracts are those in which a contractor bids for a notional quantity of specified services at times and locations to be determined by the client at a later date (for example, 100 km of resealing in district A, or 100 hours of snow removal in district B at 6 hours notice). The actual services are then "called off" in smaller packages by the client as the need arises.

6. Creating Ownership

Ownership is one of the most important building blocks in the reform agenda. How can central and local governments make road users believe that roads are their own? How can they persuade users to take an active interest in road management? This chapter tackles four related topics. First, what is meant by ownership? Second, which grass roots organizations represent road users and other interested parties, and do these provide a sound basis for involving road users in discussions on road management? Third, which institutional structures might be used to involve road users in management? Finally, how does one establish a road board, assuming that it is an appropriate way of involving road users in management of roads?

The Concept of Ownership

Ownership is a means to empower, to encourage the public to take an active role in the management of roads. For example, enthusiastic user support is a precondition for solving the problem of road financing, whether by raising taxes and reforming the budget process or by introducing an explicit road tariff. Finance ministries are always reluctant to raise taxes and user charges—the public invariably complains. In fact, the ministry of finance will likely mobilize more revenue to finance road maintenance only if road users openly express willingness to provide the extra revenues. Since road users have good reasons to see more money spent on road maintenance (see chapter 2), the issue boils down to finding ways to make known their support for a sustainable road financing plan. This mechanism for solving the financing problem also addresses other daunting management issues.

Road users may be willing to pay for roads, but only if their money is actually spent on roads and the work is executed efficiently. This important concept, that of self-interest, is part of the symbiotic relationship underlying market discipline. Road-user involvement acts as a surrogate for market discipline, encouraging the road agency to use resources efficiently and to prevent it from abusing its monopoly power. The benefits of ownership transcend financing and market discipline, however. Once road users are convinced that the government is trying to serve their needs, they will generally support a whole range of initiatives designed to improve the road sector. Ownership can become the basis for a genuine partnership between road users and the government to improve road safety, restrict fuel smuggling (or at least find an alternative the fuel levy for financing roads), and control overloading.

There are numerous examples of road agencies and road users working together to solve common problems. Together they consult about changing regulations, particularly those relating to vehicle weights and dimensions, and often collaborate on initiatives designed to improve road safety and enforce axle-weight standards (box 6.1). In some countries—Zambia being a noteworthy example—the trucking association has repeatedly put forward proposals to improve administration of international transit fees and has requested that the transport ministry allow the private sector to enforce axle-weight regulations. This system often operates in countries with extensive international transit traffic. When foreign vehicles overload and avoid paying transit fees, it undermines the international competitiveness of local hauliers. Not surprisingly, the trucking associations in transit countries like Malawi, Jordan, and Zambia have all expressed willingness to help the road agency enforce axle-weight regulations.

Box 6.1 The trucking association in Finland helps to control vehicle overloading

When the Finnish National Road Administration (FinnRA) first installed weighing-in-motion equipment, instead of using it to prosecute and fine overloaded vehicles, FinnRA decided to use it as a means of gathering information on the extent of overloading and publicizing the names of persistent offenders. At that time, FinnRA was having trouble with overloaded timber trucks using both rural roads and main trunk highways. Part of the problem was related to confusion about how the weights of timber, gravel, and peat were measured. During the 1960s and 1970s the loads were measured by volume, and the legal volumetric measure did not always coincide with the legal weight measure. FinnRA therefore approached the trucking association, which represented the timber transporters, and attempted to agree on a solution to the problem. The trucking association explained that the wood and paper industry was very competitive and that transport costs were a major factor affecting their input costs. In return, FinnRA explained how overloading affected the road pavement, particularly during the spring thaw.

Both parties agreed on an acceptable modus operandi. The Ministry of Transport and Communications agreed to permit certain overloads when the road beds were frozen in return for strictly enforced vehicle weight limits on rural roads during the spring thaw. To better define the "controlled" period, FinnRA used more than 100 freeze-depth measuring devices on the rural road network. It also prepared photographs to show what a legally loaded timber truck should look like and printed a small booklet that it distributed to its field staff and transporters for information. Because of the above discussions, the raising of the legal weight limits, and abandonment of the volumetric measurement of loads, the wood industry changed the timing of its timber purchases (to avoid the spring thaw period), and altered its inventory practices to suit. The trucking association also encouraged its members to police themselves and established a good working relationship with FinnRA's local regions who met periodically as needed.

The problem of overloading has not been eliminated, but it has been substantially reduced. The weighing-in-motion tools are still being used to gather and publicize information on overloading. The availability of this information, together with the visible damage done by the remaining overloaded vehicles has, with the assistance of the trucking association, created a form of public oversight that discourages overloading more effectively than could pure enforcement alone.

Source: Prepared by Antti Talvitie for this study.

Organizations Representing Road Users

The objective is to identify an appropriate institutional mechanism for building a public-private partnership between the politicians who represent the public (and the civil servants who assist them) and the road users. Politicians, both national and local, help to set road sector priorities, while the road users strengthen governance and provide access to private sector commercial know-how. Road users also emphasize technical considerations over narrow political interests and help to depoliticize the setting of priorities.

Road users can be easily involved through constituencies, which link the representative individual with large, assertive groups that have compelling interests in well-managed roads. Most countries possess a number of such organizations that are influential at different levels of government. These include:

• National, economy-wide organizations: chambers of commerce, farmer organizations, consultant organizations, engineering societies, pedestrian and cycling lobbies, consumer groups, and women's organizations.

• National transport sector organizations: transport institutions, transport training institutes, transport consultative councils, and transport workers unions.

• National road sector organizations: roads associations (or federations or societies), motoring organizations, trucking associations, and national organizations representing bus owners and operators.

• Local transport organizations: taxi associations and local organizations representing bus owners and operators.

• Local community organizations: village associations, parent-teacher associations, and other local community groups.

The groups most relevant for establishing ownership in the road sector are chambers of commerce, farmer organizations, engineering institutions, road associations, trucking associations, national organizations representing bus owners, other motoring groups, and labor unions. Local community organizations and taxi associations are also relevant at the local government level.

Nearly all countries have some form of business association or chamber of commerce. They are generally well-organized, take a keen interest in the state of the road network, and have a great deal of influence. Their involvement in discussions about the future of the road sector is essential. In addition, most countries have farmer organizations that are well-organized and influential, particularly when they represent large commercial farmers, a constituency that is dependent on well-managed roads. Most countries also have reasonably well-organized engineering institutions that act as opinion leaders and add a professional dimension to discussions about road management. Representatives from the engineering profession are regularly consulted about road sector issues in most countries. Finally, most countries have reasonably effective national organizations representing the trucking industry. They are regularly involved in discussions about axle-weight controls and other road-related matters.

Industrial countries also have motoring associations, often with large memberships. These associations are influential, given their size, and are regularly consulted by the government on road sector issues. Organizations representing car users and public transport operators are less common in developing and transition economies. Many of these countries have no formal mechanism for carrying on a dialogue with these potentially influential road users, cannot effec-

tively involve them in discussions on road management, or cannot work with them to confront other road sector issues. Establishing and strengthening such organizations should be an important part of any agenda for improving the management and financing of roads (table 6.1).

Ways of Involving Road Users

Road users can be involved in an advisory or executive capacity, in overall management, in management of parts of the road network (particularly at the local government level), or in specific aspects of road management. Most countries invite outsiders to join steering committees that guide consultants working on the road sector, or to sit on specialized advisory boards that review departmental research programs, training programs, road design standards, and other technical matters. For example, in England there is a Road Users Committee that facilitates dialogue between the Highways Agency and representatives of both motorized and nonmotorized road users.

Several countries have national road safety councils that include representatives from a wide variety of private sector organizations (table 6.2). The councils attempt to coordinate the activities of different organizations involved in road safety and may also advise the transport ministry on a wide range of matters related to road safety. Most councils lack statutory powers, are underfund-

Table 6.1 Organizations representing road users in selected countries

Country	Road associations[a]	Trucking organizations	Bus owners and operators	Motoring associations	Engineering professions
Argentina	Yes	Yes	Yes	Yes	Yes
Chile	Yes	Yes	Yes	Yes	Yes
Finland	Yes	Yes	Yes	Yes	Yes
Germany	Yes	Yes	Yes	Yes	Yes
Ghana	No	Yes	Yes	No	Yes
Hungary	Yes	Yes	No	Yes	Yes
Japan	Yes	Yes	Yes	Yes	Yes
Jordan	Yes	Yes	No	Yes	Yes
Korea, Rep. of	No	Yes	Yes	Yes	Yes
New Zealand	Yes	Yes	Yes	Yes	Yes
Pakistan	No	Yes	Yes	No	Yes
Russia	No	Yes	No	Yes	No
South Africa	Yes	Yes	Yes	Yes	Yes
United Kingdom	Yes	Yes	Yes	Yes	Yes
United States	Yes	Yes	Yes	Yes	Yes

a. Road associations, federations, societies, and road engineering associations mainly represent consultants, contractors, and plant and materials suppliers.

Table 6.2 Institutions involving road users in management of roads

Country	National road board or council	Advisory committees	National road safety council commission or similar	Local road boards/ cooperatives
Argentina	No	No	Yes[a]	Yes
Chile	No	Yes[b]	Yes	Yes
Finland	Yes	No[c]	Yes	Yes
Germany	No	Yes	No	No
Ghana	Yes	No	Yes	No
Hungary	No	No	Yes	No
Japan	Yes	Yes	Yes	No
Jordan	No	No	No	No
Korea, Rep. of	No	No	Yes	No
New Zealand	Yes	Yes	Yes	No
Pakistan	Yes	Yes	No	No
Russia	No	Yes	Yes	No
South Africa	Yes	Yes	Yes	Yes
United Kingdom	Yes	Yes	No	No
United States	No	Yes	Yes	Yes

a. In the process of being established.

b. Advises on pavement management systems.

c. Road plans are also circulated to many interest groups for comment.

ed, and do not have an effective secretariat, though some do function well, serving as a useful body for involving the private sector in discussions on road safety.

Urban and rural district councils are not particularly good at involving road users in the management of roads. The usual mechanism for doing so is through working committees that operate at the local government level. All urban and rural district councils have committees that handle finance, planning and development, housing, and the other functions delegated from higher government levels. Some also have road and road transport committees that deal with roads, street cleaning, street lights, drainage, public transport, and traffic management. These committees generally consist of elected members and occasionally include some nonelected people (such as the police). They rarely include nonelected people representing road users or the local community. But representatives of such organizations are sometimes invited to participate in the business of the committees.

At the national and regional level road users generally participate in the management of roads through road management boards. These are becoming increasingly common—there are now active road boards in several countries, including Australia (the states of New South Wales, Queensland, and Victoria), Finland, Ghana, India, Japan, Latvia, New Zealand, South Africa, Sweden, the United Kingdom, Yemen, and

Zambia. Some of these are executive boards that manage the main road network, such as the boards of FinnRA, Ghana Highway Authority, Transit New Zealand Authority, and the South African Roads Board; others manage the road fund, such as the Ghana Road Fund Board, the board of the Malawi National Roads Authority, the board of Transfund New Zealand, the Yemen Road Fund Board, and the Zambia National Roads Board. Still others merely advise the appropriate minister on road management and financing, such as the Japan Road Council, the advisory board of the Latvian road fund, and the U.K. Highway Agency Advisory Board. The South African Roads Board— originally established in 1935—is one of the oldest boards, followed by the Japan Road Council (1952), and the original New Zealand Roads Board (1953), which then became the Transit New Zealand Authority (1989) and is now Transfund New Zealand (1996).

The South African Roads Board has an interesting history. First established in 1935, it started off with six members, four representing the provinces and two appointed by the Minister of Interior. Although the Board was meant to function autonomously with the provincial representatives acting "in the national interest," it quickly lapsed into gridlock because the provinces expected their representatives only to promote their own local interests. In 1948 the Board was therefore replaced by another composed exclusively of civil servants. This worked better, although it led to a large and controversial freeway program and to the accumulation of a large surplus in the road fund—which led to the suspension of the fuel levy in 1988.

Following the suspension of the fuel levy, the board was expanded to include representatives of local government, the engineering profession, road users, and industry and commerce. This board functioned well, initiating a successful toll road program. In fact, the public-private board worked so well that in 1995 membership was further widened to include more private sector representatives and a representative from academia. In the end the board consisted of three members representing central government, three representing local government, five representing the private sector, and one representing academia. In March 1998 the board will be replaced by an autonomous board with ten members, nine representing the private sector and one representing the Department of Finance.

The major lessons learned from the experience of the South African Roads Board are:
• It is better to have one person representing all local governments than to have each local government represented on the board.
• A board made up wholly of civil servants does not necessarily serve the interests of road users.
• A broad-based public-private board acts more commercially and better serves the long-term interests of the government and road users.

Setting up a Road Board

Several issues arise in establishing a road board: legal procedures for setting up the board; membership and procedures for nominating board members and appointing the chairperson; the role of committees, if any; the size and composition of the secretariat; and the functions of the board and its detailed terms of reference, including the relationship between the board and the parent ministry and how the board will be held accountable.

Legal Basis for a Road Board

There are three ways to establish a road board: under existing legislation, under a ministerial or presidential decree (or the equivalent), or under new legislation. The Zambia National Roads Board and the U.K. Highways Agency Advisory Board were both set up under existing legislation. The road legislation in many former British colonies gives the responsible minister power to establish a road board by publishing a notice in the gazette. Legislation differs widely and provides for establishment of advisory boards, as in Tanzania, and for executive boards, as in Zambia. The U.K. Highways Agency Advisory Board was set up under the 1994 Government Framework Document. The Advisory Committee to the National Highways Authority of India is slightly different in that the Chairman of the Authority simply decided to set up a board to advise him on all matters pertaining to managing the national highway network.[1]

The boards set up under decrees can be either advisory or executive. The board in Finland (established by parliamentary decree) is an executive board that manages FinnRA; the Yemen Road Fund Board (established by presidential decree) is an executive board that manages the road fund; the Jordan Road Fund Board (established by cabinet decree) advises the responsible minister on the management and financing of road maintenance.

Other boards have been set up under special-purpose legislation. Most of these have executive powers and manage part of the road network, the road fund, or both. The Japan Road Council is unusual in that it was specifically set up under legislation to advise the Minister of Construction on overall road policy. The newly established Malawi National Roads Administration is even more unusual. The Administration separates itself into two subcommittees—one manages the road fund in an executive capacity, while the other advises the responsible minister on the planning of road works and ways to strengthen road management. The expectation is that the advisory committee will eventually evolve into the executive board of a semi-autonomous highway authority.

Although new legislation offers the best long-term solution, it has some short-term disadvantages. First, it requires parliamentary approval, and that takes time. Second, it requires formalizing a number of operating procedures without the benefit of hindsight. This shortcoming has created major problems in countries like Mozambique and Yemen, where the original wording of the legislation made it difficult to add private sector representatives to the board, except as unpaid, nonvoting advisors. It is often better to establish the board under existing legislation or under a simple ministerial or presidential decree (or the equivalent), and to then work with provisional operating procedures before legalizing the arrangement.

The Structure of the Board

There are at least three structural factors that affect the workings of the board: its overall composition, whether it is made up of professionals or representatives of key constituencies, and the way in which members are nominated.

Who should sit on the road board? Some of the early road boards, including the 1948 South African Roads Board, the initial U.K. board, and the Administrative Council in Mozambique, were made up exclusively of civil servants. The board of the Swedish National Road Administration even included three politicians and three elected staff representatives.[2] It is difficult for such boards to build up public support, to deal with their relationship with the responsible minister, and to make difficult decisions that might appear critical of ministers and civil servants. The boards' proceedings are often confidential (the U.K. board is covered by the Official Secrets Act), and they tend to spend too much time dealing with day-to-day administrative matters rather than important long-term policy issues.

The boards that do include a wide range of road sector interests—as in Finland, Japan, Latvia, South Africa, and Zambia—tend to be more effective. The same is true of the boards in New Zealand. The board of Transfund is made up of sector representatives, while the board of Transit New Zealand is made up of prominent people selected for their skills and professional standing. These boards are innovative, act decisively, make great efforts to consult the public, do not hesitate to be critical when needed, and generally operate in a commercial manner. A representative road board, which includes representatives of road users, is therefore preferable to one made up wholly of civil servants or one with a majority of civil servants.

A broad-based, representative board must comprise people with strong vested interests in having well-man-

aged roads. Some typically represent the central government departments directly concerned with roads—usually the ministries of finance, works, transport, and agriculture. Others comprise nongovernmental members representing road users—usually the business community, the road transport industry, farming interests, and the engineering, accounting, or legal professions. Members may also represent local governments. The nongovernment members represent the people who use the road network and also pay for it, while the local government members represent agencies that depend heavily on roads for delivering public services to their constituents. They thus have a strong vested interest in ensuring that the road network is managed efficiently and effectively.

Should the board be made up of professionals or people representing key constituencies? The boards in Japan and New Zealand (Transit New Zealand) are unusual in that their boards are made up primarily of prominent persons selected for their skills and professional standing. The nongovernment members in the United Kingdom are selected on the same basis. Although these boards function well, most other countries, particularly those with problematic governance, have chosen a more explicit link between nongovernment board members and their constituencies. Each member represents a specific constituency, like the chamber of commerce, national farmers association, road haulage association, institute of engineers, or association of municipalities. The Sierra Leone Road Authority is one example of a balanced tripartite board structure. One-third of its members represent government departments, one-third represent road users (the chamber of commerce, the road transport industry, and the engineering profession), and the final third represent regional interests (local government).

Flexibility in the size of the board allows the composition to evolve in line with changing road needs and provides a useful vehicle for resolving conflicts over membership. For example, the inaugural meeting of the Zambia National Roads Board was delayed for several months because of a disagreement over which of the two main road transport organizations should represent the road transport industry. This conflict could have been avoided had it been possible to appoint board members to undesignated positions on the "advice of the minister" or on the "recommendation of the board." The boards in Ghana, Jordan, Malawi, and New Zealand (Transfund) all provide for appointment of ad hoc members representing the public interest based on the recommendation of the board, permanent secretary, or road transport interests (box 6.2).

It is difficult to say how and how well these road boards function, since many are unduly secretive and unwilling to open their meetings to the public or to share their agendas and minutes with outsiders. But based on what is known publicly about them, we can draw some tentative conclusions. Boards that consist solely, or primarily, of civil servants—particularly when the permanent secretary or minister is chairperson—tend to avoid controversial issues, are reluctant to criticize the road agency, and spend little time discussing overall strategic issues. Representative boards, on the other hand, particularly those with independent chairpersons, function better. Even if the responsible minister serves as chair, the private sector members encourage more open and constructive debate. Further, strong chairpersons avoid calling for votes on controversial issues—they try to manage by consensus. The same applies to disagreements between the board and the minister. Most chairpersons work hard to avoid such confrontations and regularly consult both the minister and their constituents about the business of the board. Therefore, major conflicts between a representative board and the minister are rare.

How should board members be nominated? Members must be able to speak on behalf of their constituents and report back to them. Government departments must be represented by the permanent secretary or by a senior official with regular access to the permanent secretary. Otherwise, the member will not really represent the ministry. Likewise, NGOs cannot be represented simply by an acquaintance of the minister who happens to run a trucking company. Such people may have no mandate from the road transport industry. They have no constituency and no way of communicating with or mobilizing the support of road users. Genuine ownership occurs only when the people selected to represent each constituency truly represent and can consult with their members.

Several countries have attempted to address this concern by specifying that no ministry may be represented on a board by anyone below the level of director. In the case of nongovernment members the organizations represented on the board are either permitted to nominate their own representatives, or the responsible minister invites them to nominate candidates and then appoints one of the nominees to the board. New Zealand, some Australian states,[3] and South Africa consult the organizations about their nominees, while Ghana, Jordan, Latvia, Malawi, Zambia, and several other countries let the organizations nominate their own members. The United Kingdom is unusual in this respect. Civil servants prepare a list of nominations which are then discussed with the CEO of the Highways Agency. The permanent secretary then selects the members and informs the minister of his choice. Once nominated, board members are generally appointed by the minister, president, or cabinet. In many cases the names of members are published in the government gazette.

There is also the important question of how to choose the chairperson. There are four broad options. The chairperson can be ex officio; nominated and appointed by the minister, perhaps after consulting the board; appointed by the minister from among the existing members of the board; or elected from among the board members. Finland and South Africa both have ex officio chairpersons. The chairperson is the director general of roads or transport. Such an arrangement often creates a conflict of interest. The head of the board will often find that he/she is criticizing the agency that he/she directs. Furthermore, when the director general is chairperson, outsiders tend to view the board as a lobby group arguing on behalf of the road agency rather than as an impartial body concerned only with the national interest. It is worth noting that South Africa has amended its legislation to appoint an independent chairperson to head their new Roads Agency.

Usually, the responsible minister nominates the chairperson, as in Japan and Ghana (Ghana Highway Authority). This method works well provided the minister chooses a competent person and ensures that the nominee is able to work well with the other members of the board. Another option designed to avoid con-

Box 6.2 Membership of road boards in Japan, New Zealand, Finland, the United Kingdom, and South Africa

JAPAN ROAD COUNCIL: The Council was established in 1952 and consists of a chairperson and 12 members. Members are nominated by the director general of roads and appointed by the minister of construction. The chairperson has traditionally been president of the Japan Road Association (always a former undersecretary from the Ministry of Construction), but is currently former president and chairperson of Nissan Corporation. Board members include representatives of the motor industry, business community, trade unions, academia, and local government. Much of the Council's substantive work is carried out by three subcommittees: one deals with road policy, one with toll roads, and the other with environmental issues. The Council has no permanent secretariat.

TRANSIT NEW ZEALAND AUTHORITY: The original National Roads Board was established in 1954 with the minister of works as chairperson. The current board was established in 1989 and consists of a chairperson, deputy chairperson, and six members. Members are appointed by the governor-general on the joint recommendation of the ministers of transport and finance and based on consultations with various industry associations. The chairperson is appointed by the governor-general from among the members of the Board. The current chairperson is a former local authority engineer. The deputy chairperson is past president of the Institution of Professional Engineers, and the other six members have backgrounds in town planning, industry, local government, road transport, farming, and accounting. The Board sets up ad hoc subcommittees to deal with specific tasks. The subcommittees always include the chairperson and usually two other board members. The Corporate Services Manager of Transit New Zealand acts as secretary to the board, and the secretarial functions absorb about one person-year per year.

BOARD OF FINNISH NATIONAL ROAD ADMINISTRATION (FINNRA): This Board was established in 1990 and consists of a chairperson and seven members. Members are nominated by the director general of FinnRA and appointed by the cabinet on the recommendation of the Minister of Transport and Communications. The director general is ex officio chairperson, and the other seven members currently represent the Ministries of Transport and Environment, the municipalities, the Confederation of Industries, the road transport industry, the labor union, and the union of employees. The Board has no subcom-

mittees. One of the directors of FinnRA handles its business, and a secretary takes minutes and arranges meetings. The secretarial functions absorb about one person-year per year.

U.K. HIGHWAY AGENCY ADVISORY BOARD: Established in 1994, the U.K. Board consists of a chairperson and four or more members. Members are appointed by the permanent secretary, after consultation with the Agency's chief executive officer, based on nominations prepared by staff of the Department of Transport. The minister is informed of the appointments. The permanent secretary is ex officio chairperson, and the Board does not meet in his/her absence. The other members include the chief executive, one or more of the Agency's executive Board members, one or more departmental representatives, and one or more nondepartmental members. There are currently two nondepartmental members who are appointed in their personal capacity. The Board has no subcommittees and no permanent secretariat. One of the department's staff acts as secretary to the Board and is assisted by other departmental staff. The secretarial functions absorb about five person-weeks per year.

SOUTH AFRICAN ROADS BOARD: The original Board was established in 1935. The Board operating until 1998 was established in 1988 and consisted of 12 members—6 public and 6 private—appointed by the minister of transport, posts, and telecommunications, following consultation with the constituencies represented on the Board. The director general, transport, was ex officio chairperson. Two members represented the central government, two represented provincial governments, and one represented metropolitan and local authorities. The six private sector members represented car users, the bus industry, the road freight industry, business (commerce, industry, mining, and agriculture), the engineering profession, and academia. The Board had three committees: finance, tenders, and urban transport and planning. The urban transport and planning committee managed the Urban Transport Fund, reviewed the transport plans prepared by the core cities, and made recommendations on these plans to the Board. The Board was terminated in 1998 and its powers devolved to the provinces. Until 1995 there was also a toll road committee, which advised the Board on all matters pertaining to toll roads. The chief director, roads, acts as secretary to the board. The secretarial functions absorb about two to three person-years per year.

flicts between the chairperson and the board requires that the minister appoint the chairperson only following consultation with the board. This mode has been adopted recently in Japan. A third option leaves the choice of chairperson to the minister, but limits potential candidates to serving members of the board. This method is used in New Zealand (Transfund and Transit New Zealand) and in Malawi. The option in which the board elects its own chairperson is rare. Both Jordan and Zambia use this option. It works well in Zambia, and several other African countries have shown interest in the procedure.

Subcommittees and the Secretariat

Several road boards have standing committees to carry out much of the substantive business. Some boards invite technical specialists to sit on these committees. In Japan standing committees carry out most of the work of the Road Council. In Malawi and South Africa the road fund is managed by a committee of the board. The committees in Japan deal with toll roads, environmental matters, road policy, and road financing. Those in South Africa manage the road fund and the urban transport fund (the functions of the urban transport committee were devolved to the provinces in 1998), while those in Malawi manage the road fund and advise the responsible minister on road management. Similarly, in Zambia a technical committee—made up primarily of technical specialists—reviews the road maintenance proposals put forward by the district councils, examines the draft contract documents, reviews the bids and bid prices, and then advises the main board on whether the proposals should be financed "as is" or amended.

Most boards have some form of secretariat. The Japan Road Council is unusual in that it has no independent secretariat, which is considered to be a major disadvantage. The duties of the secretariat include organizing the meetings of the board, taking minutes, handling correspondence, keeping a record of accounts, and preparing position papers at the request of the board. The amount of time spent on these activities varies from a few weeks per year (the United Kingdom) to a full-time job for one or more people (Finland, South Africa, Transit New Zealand, Transfund New Zealand, and Zambia). The amount of time required clearly depends on the scope of the board's activities and whether it is new or well-established. An effective board needs some kind of secretarial capacity, and the board will be more effective if the secretariat is independent of the agency it is overseeing.

Functions and Detailed Terms of Reference

The terms of reference for the board are usually spelled out in the legislation or in the legal regulations promulgating the legislation. The terms must cover the relationship between the board and the parent ministry, whether the board is executive or advisory, its sources of finance, and its day-to-day responsibilities. Functions vary widely from country to country (boxes 6.3 and 6.4). Legislated duties are usually supplemented by additional duties and procedures, which may be specified by the responsible minister (New Zealand) or decided by the chairperson (Finland). To prevent unwarranted interference in the board's business, the legislation may also define the circumstances under which the minister can override board decisions and specify the way in which the minister must issue directions to them. Such directions generally have to be specified in writing and may also be limited to matters already covered in the legislation.

The legal regulations are detailed, fairly standard, and usually cover the appointment of board members, their tenure, payment of fees and expenses, secretarial arrangements, frequency of meetings, keeping of minutes, accounting arrangements, submission of reports and their content, and auditing arrangements. In the case of a road fund board the regulations (or primary legislation) may also include special penalties relating to misappropriation of funds. Reporting arrangements are particularly important since they act as a vehicle for holding the board accountable for its work. They keep the parent ministry informed, enable board members to report back to their constituents, and also help keep the public informed.

The board is usually required to submit an annual budget, an annual statement of accounts (audited by the auditor general or independent auditors appointed by the auditor general), and an annual report that includes information on board policies and activities during the year. The boards of the National Roads Authority in Malawi and Transfund New Zealand are also required

---

Box 6.3 Duties laid down for road boards in Japan and New Zealand

JAPAN ROAD COUNCIL: The role and duties of the Council are laid down in article 77 of the Road Law, 1952. The law provides for a Road Council to be established by the Ministry of Construction at the request of the minister. The Council's role is to:

• Investigate present road conditions and propose future improvements.

• Deliberate on management of the road fund and toll-road financing, and advise the minister on changes necessary to reorient road financing.

• Examine important topics like road safety, traffic congestion, and environmental damage, and propose a long-term strategy on road policy to be adopted by the government just before the start of the five-year road improvement programs.

• Deliberate on the contents of the five-year road improvement programs prepared by Ministry of Construction and, after the Council is satisfied with the program, convey that consent to the minister.

TRANSIT NEW ZEALAND AUTHORITY: The National Roads Board was established in 1954 and the legislation was amended in 1979, 1989, and 1996. The present Board established under the Transit New Zealand Act, 1989, was empowered to:

• Prepare an annual national land transport program and review and revise the program from time to time.

• Manage implementation of the following elements of the program: local roading, safety (construction and maintenance), passenger transport, state highways, and administration.

• Make payments from the [road fund] account and, in special cases, present proposals to the minister for funding outside the approved national land transport program.

• Control the state highway system, including planning, design, supervision, construction, and maintenance.

• Advise local authorities in relation to their functions, duties, and powers and audit the performance of every local authority as compared with its statement of intent contained in the relevant land transport program.

• Provide the minister with such information and advice as the minister may require and carry out such other land transport functions and duties as the minister may prescribe.

---

to prepare a consolidated national land transport program that incorporates the programs of the regional transport authorities. The road fund boards likewise have to publish quarterly financial reports, together with the audited annual accounts of the road fund.

Key Recommendations and Conclusions

The key issues that have emerged from the above discussion are as follows:

• The central theme of this chapter is ownership—which is seen as a means to empower road users and encourage them to take an active interest in the management of roads. Among other things, their involvement creates a form of surrogate market discipline that can encourage the road agency to use resources efficiently and prevent it from abusing its monopoly power.

• The overall objective is to build a public-private partnership between the politicians who represent the public as a whole (and the civil servants who assist them) and the road users who have a strong vested interest in well-managed roads. This pairing helps to depoliticize the setting of priorities, since road users tend to emphasize technical considerations over narrow political interests.

• The challenge is to identify an appropriate institutional mechanism that can effectively involve politicians and road users in objective, depoliticized discussions about road management. This typically cannot be done by simply involving them as individuals. Effective involvement generally requires constituencies that link the representative individual to large, assertive groups with compelling interests in well-managed roads.

• The organizations that are most relevant for establishing ownership in the road sector are chambers of commerce, farmer organizations, engineering institutions, road associations, trucking associations, national organizations representing bus owners, other motoring groups, and labor unions. Local community organizations and taxi associations are also relevant at the local government level.

• Although road users can be usefully involved by way of more or less informal committees and consultative councils, at the national and regional level road users are increasingly participating in discussions on road

---

Box 6.4 Duties laid down for road boards in Finland, the United Kingdom, and South Africa

BOARD OF FINNISH NATIONAL ROAD ADMINISTRATION: This Board was established under the Finnish National Road Administration Decree, 1990. It was set up to involve the public in discussions on road sector development, strengthen concern for the environment, improve the effectiveness and efficiency of FinnRA, and make road planning more transparent. The decree empowers the Board to:
• Make decisions on FinnRA's goals and operations, taking into account the goals set by the Ministry of Transport.
• Decide on administrative arrangements.
• Decide on the budget proposal, the activity and financial plans, and long-term development programs.
• Monitor implementation of the organization's goals and approve the financial accounts.
• Decide on important research and development tasks.
• Issue the organization's rules and regulations, except where these functions have been delegated to the director general or other functionary.
• Deal with any other matters of importance to the organization as decided by the chairperson.

U.K. HIGHWAYS AGENCY ADVISORY BOARD: This was established as a nonstatutory board in 1994 through publication of a Government Framework Document. The Board's general role is to support the permanent secretary in advising the secretary of state on the strategic direction of the Highways Agency. In particular, the board advises on the Agency's:
• Corporate and business plans.
• Performance against the targets set in its corporate and business plans.
The board also supports the Agency's chief executive to achieve his/her aims and objectives.

SOUTH AFRICAN ROADS BOARD: The original National Road Board was established in 1935. The current South African Roads Board was set up in 1988, and the legislation was amended in 1995. The main purpose of the Board is to promote and encourage the development of transport and, where necessary, to coordinate various phases of transport in order to achieve the maximum benefit and economy of transport services to the public. The main objectives of the Board are to:
• Design, build, and maintain a national network of freeways and other roads, including toll roads.
• Compile a priority list of roads to be built or improved.
• Design and build various special roads that are in the national interest.
• Set geometric standards for the construction of national and special roads.
• Preserve the environment.
• Expend available funds in the most cost-effective manner in providing a primary road network.
• Initiate research, whether in South Africa or elsewhere, in connection with the design, planning, or construction of roads.
• Grant bursaries or subsidies to enable people to study or research any subject connected with roads.
• Advise the minister, at his/her request, on questions relating to roads that may be raised by the government of any other country or territory.
• Provide rest and service areas, in conjunction with private enterprise, at strategic points on national roads in order to promote road safety.
The Department of Transport is charged with carrying out the executive and administrative work necessary to enable the board to carry out the duties and functions assigned to it.

---

management through road management boards. These are now widespread in industrial, developing, and transition economies.
• One of the first issues that arises when setting up a road board is whether it should have executive powers or simply operate in an advisory capacity. If it is an executive board, it will usually require new legislation. Otherwise, it can often be established under existing legislation using simpler parliamentary procedures or using a ministerial or presidential decree (or the equivalent). In the long term an advisory board will be on firmer ground if it is supported by legislation.
• To be effective, the board must have representative membership. In addition to civil servants representing

key government ministries, it should also include people representing road users, the business community, farmers, the professions, and local governments. The membership will vary depending on the functions of the board.
• Board members should ideally represent constituencies that have strong vested interests in well-managed roads, not simply individuals without clearly defined constituencies. The represented constituencies should nominate their own members. Civil servants representing government ministries should not be below the level of director. Board members should be formally appointed by the responsible minister, president, or cabinet.

- It is advisable to have one or two ad hoc members nominated by the board or the minister to represent the public interest. This provision will allow for some flexibility in membership.
- The chairperson should be independent and should ideally be chosen from among existing board members (elected by the board or chosen and appointed by the minister), or appointed by the minister following consultation with the board.
- The board should be able to establish standing committees that can invite technical specialists to participate in their meetings. This ability gives the board access to a wide range of technical knowledge. The board should also have an independent secretariat to organize meetings and assist with its business.
- The board should have clear terms of reference that can be supplemented and updated by the responsible minister. Among other things, the terms of reference should spell out the relationship between the board and the parent ministry.

- The board's rules and procedures must be clearly spelled out in legal regulations. The regulations should include procedures for appointing board members, their tenure, frequency of board meetings, auditing arrangements, and the content and timing of annual reports.

Notes

1. India has a 20-member advisory committee that advises the Chairman of the National Highway Authority. The members represent the road transport industry, business (including vehicle manufacturers), research and academia, transport training, and local government.
2. The board consists of three politicians (one is the chairperson), the director general, a civil servant, a contractor, an NGO, a representative from academia, and three elected staff representatives.
3. The Roads Corporation Advisory Board in Victoria has formalized the appointing procedure by inviting the various constituencies to submit a panel of three names for the minister's consideration.

7. Ensuring an Adequate and Stable Flow of Funds

This chapter examines pricing and cost recovery policies for roads, developing a model that attempts to promote economic efficiency and to generate sufficient revenues to operate and maintain the road network on a sustainable long-term basis.[1] To do so, the model must influence the demand for travel—whether and how to make the journey—as well as the supply of road services. The impact on supply is particularly important. The road agency should be discouraged from simply passing on to road users the costs of its own inefficiencies in the form of higher user charges.[2] Instead, financing mechanisms should encourage the road agency to use resources efficiently, limit the scope of the road network to what is affordable, and construct new roads only when resources are available for maintenance. In other words, pricing and cost recovery policies should bring roads into the marketplace by setting a clear price and subjecting the road agency to a hard budget constraint, by linking revenues and expenditures, to promote surrogate market discipline.

This chapter addresses two key questions: Which instruments can be used to charge road users? Which principles should guide the pricing and cost recovery policies that are applied to roads?

Setting Clear Market Signals

To influence demand and provide a basis for linking revenues and expenditures, charging instruments should be:
- Easily recognizable.
- Related to road use.
- Easy to separate from indirect taxes and other service charges or fees.

- Simple to administer, that is, not vulnerable to widespread evasion, avoidance, and leakage.

In addition, the instruments should be able to distinguish between paying for the right to use the road network, actual travel on the roads, the occupying of road space (either by parking or causing congestion), and the benefits of road access.

Selecting Appropriate Charging Instruments

The main instruments used to charge road users include vehicle license fees, levies or taxes on transport fuels, international transit fees, and tolls. Very few countries use supplementary heavy-vehicle fees, although many express interest in doing so. Parking charges are common in urban areas, whereas weight-distance fees and the various methods of charging for urban road congestion are rarely used. Although some countries deposit certain sales and excise taxes into their road funds, these are nearly always general revenue taxes and should not constitute part of road-user charges. In other words sales and excise taxes on transport-related services are generally set at the same rate as they are for all other comparable goods and services (see box 3.1). Likewise, some countries wrongly treat drivers licenses, vehicle inspection fees, and registration fees like road-user charges. These are nearly always service fees levied in connection with the provision of specific services.[3]

The instruments best suited to developing and transition economies are vehicle license fees, supplementary heavy vehicle license fees, fuel levies, and international transit fees (table 7.1). Parking charges, as they are presently collected, are less suitable because they are difficult to administer. They suffer from high levels of avoidance and leakage.[4] However, if they are collected under contract, parking charges could play an

Table 7.1 Administrative characteristics of different road-user charging instruments

Charging instrument	Potential role	Related to road use	Separable from general taxes	Easily recognizable	Collection cost (percent)	Avoidance or evasion	Ease of collecting by contract	Suitability[a]
					Administrative characteristics			
Tolls	user fee	yes	yes	excellent	10–20	moderate	simple	moderate
Vehicle license fee	vehicle access fee	no	yes	good	10–12	high	moderate	high
Heavy vehicle license fee	vehicle access fee	not directly	yes	good	unknown	unknown	simple	high
Fuel levy	user fee	partly	can be	good	negligible	low	simple	high
Weight-distance fee[b]	user fee	yes	yes	excellent	5	moderate	moderate	low
International transit fee	foreign user fee	should be	yes	good	10	high	simple	high
Parking charges[c]	control access	partly	yes	good	over 50	high	simple	moderate
Cordon charge[d]	congestion charge	partly	yes	moderate	10–15	unknown	simple	moderate
Area license	congestion charge	partly	yes	moderate	10–15	unknown	simple	moderate
Electronic road pricing	user or congestion charge	can be	yes	good	less than 10	unknown	simple	low

a. Suitability as general charging instruments.
b. A simpler form of weight-distance fee is the vehicle-km fee. It employs the same basic principles, but relates fees more simply to vehicle type and distance.
c. These are difficult to administer and currently generate little revenue.
d. These are only suitable when the road network lends itself to cordon pricing.
Source: Based on case studies in Argentina, Bolivia, Ghana, India, Tanzania, Zambia and Yugoslavia. See Heggie (1992).

important role in helping to generate revenues and to manage urban traffic.

The other technically sound charging instruments, tolls and weight-distance fees, are less suitable as general charging instruments. Few roads carry sufficient traffic to make widespread tolling economic, and weight-distance fees are difficult to administer in developing and transition economies (see box 7.1). The advantage of weight-distance fees is that they encourage the use of vehicles with axle configurations that do less damage to the road pavement. They also make it easier to charge for roads when there is rampant fuel smuggling and, particularly if introduced regionally, make it easier to charge international truck traffic. Weight-distance fees should therefore be considered as soon as a country has developed the capacity to administer them.

Most countries use vehicle license fees (usually based on gross vehicle weights or engine capacity), a few use license fees based on axle weights or vehicle weight, a surprising number use fuel levies, and some also use international transit fees (table 7.2). Several countries also use tolls on bridges, ferries, and selected high-volume roads. These charges may be used as a two-part road tariff. The license fees can be used to charge for access to the road network, while the fuel levies, international transit fees, and tolls can be used to charge for use of the road network. Fuel consumption is not exactly related to variable road maintenance costs, but is related closely enough for practical charging purposes (see figure 7.1). In terms of revenues raised, fuel levies are by far the most important user charges currently used (see figure 7.2).

Administrative Considerations

It is important to ensure that the above fees, fuel levies, and bridge, ferry, and road tolls are administered efficiently. This means minimizing evasion, avoidance, and leakage; avoiding inadvertent subsidies; ensuring that the fuel levy does not unintentionally tax non-transport users of diesel; and minimizing fuel price distortions.

In developing countries license fees suffer from widespread evasion, international transit fees suffer from serious leakage (aggravated by the fact that they are often paid in foreign exchange), and tolls can be costly to administer and—particularly in the case of ferry and bridge tolls—suffer from high levels of evasion and leakage (table 7.1). In some countries half of the vehicles go unlicensed and uninsured, revenues from international transit fees are less than half their potential, and the costs of administering ferry and bridge tolls is higher than the revenues collected.

Box 7.1 Weight-distance fees for diesel vehicles

New Zealand and Iceland use weight-distance fees to charge diesel vehicles for road usage. Norway and Sweden used weight-distance fees until the early 1990s, but have now abolished them. Namibia is planning to introduce them in the near future, and a weight-distance fee, the Euro Vignette, is under consideration in the European Union for charging foreign vehicles traveling through member countries (for example, Russian trucks entering Finland). The basic principle is that all diesel vehicles must buy a license (in New Zealand they are issued in multiples of 1,000 km) graduated according to axle configuration and gross vehicle weight. The charges are administered through sealed hub odometers or other certified distance meters. The charge is lower for vehicles with multiple axles and increases with gross vehicle weight (see figures below).

The weight-distance fee is administered separately from the general tax system, and all revenues collected from the sale of weight-distance licenses are paid into a special account set aside to support spending on roads. In Norway

and Sweden revenues were not paid into a special account. In addition to the weight-distance fees, Iceland and New Zealand also levy a special charge on gasoline. The revenues from this charge are also paid into the special account.

Weight-distance fees can be difficult to administer. There is considerable scope for evasion—mainly by understating vehicle weight—unless the sale of licenses can be checked for consistency and linked to an active enforcement program. In New Zealand it is estimated that collection and enforcement absorb 3.2 and 2.0 percent respectively of gross revenues, evasion accounts for about 12 percent of net revenues (9.4 percent from heavy vehicles and 2.8 percent from light vehicles), and legal avoidance for 7 percent of net revenues. The system works satisfactorily when it is effectively administered—with fees collected under contract—and vigorously enforced. But the collection technology is now somewhat dated and countries should perhaps wait until electronic systems are available before introducing weight-distance fees.

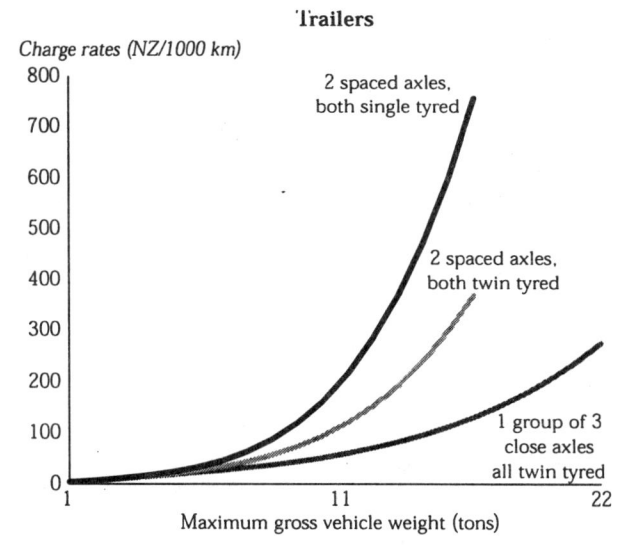

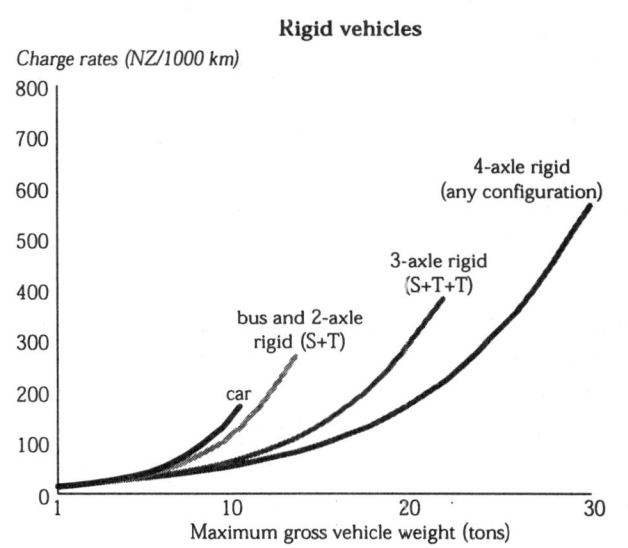

Governments are attempting to improve revenue administration to lessen these problems.

There are two main options. The first is to simplify the fee structure, which in turn will simplify administration and reduce the costs of collection and compliance. One way of doing this is by mobilizing most license fee revenues through a supplementary heavy-vehicle license fee. There are fewer heavy vehicles (perhaps 20 percent of the total vehicle fleet), and they are mostly owned by registered businesses. Thus a heavy-vehicle license fee would be easier to administer. Some

countries already have such fees, and others are considering them. The second option is to collect more fees under contract with the private sector. Countries as diverse as New Zealand, Yemen, and Mozambique have done so with good results (box 7.2).

Administrative arrangements may also lead to inadvertent subsidies. Vehicles owned by the government rarely pay license fees, and government and diplomatic vehicles often pay no fuel levies. These vehicles nevertheless impose measurable costs on the road network, and typically other users have to pay these costs.

Table 7.2 Charging instruments currently used in selected countries

Country	License fees	Supplementary heavy-vehicle fee	Fuel levy	International transit fee	Other charges
Argentina	Yes	No	Yes[a]	No	Fuel tax and tolls
Chile	Yes	No	No	No	Tolls
Finland	Yes	No	No	Yes	Fuel tax
Germany	Yes	No	No	No	Fuel tax
Ghana	No	No	Yes	No	Tolls and vehicle inspection fees. The fuel levy applies to all fuels
Honduras	Yes	No	Yes	No	Tolls
Hungary	No	Yes[b]	Yes	Yes	None
Japan	No	Yes	Yes	No	Vehicle purchase tax at local level.
Jordan	Yes	No	No	Yes	Gasoline tax earmarked to municipalities
Korea, Rep. of	Yes	No	Yes	No	Excise taxes on automobiles; tolls
Latvia	Yes	No	Yes	No	None
New Zealand	Yes	No	Yes	No	Weight-distance charges
Pakistan	Yes	Yes	Yes	No	Tolls; vehicle registration and vehicle sales tax
Russia	Yes[c]	No	Yes	No	Vehicle sales tax; enterprise taxes in selected cities
South Africa	Yes	No	Yes[d]	Yes	Fuel tax
United Kingdom	Yes	No	No	No	Fuel tax
United States	Yes	Yes	Yes	No	Graduated tax on tires; retail tax on selected trucks and trailers
Zambia	Yes	No	Yes	Yes	None

a. A portion of fuel taxes allocated to provinces is earmarked on a cost-share basis.
b. Weight-related vehicle tax based on gross vehicle weight.
c. For vehicles registered in Moscow and St. Petersburg.
d. Reinstated as of March 1998.

To avoid the distortions that such exemptions create, all road users should pay license fees and fuel levies, or the government should reimburse the road agency for loss of revenue caused by exemptions. An implicit subsidy is also given when the pump price of fuel (excluding the fuel levy) is lower than its border price (box 7.3). When that occurs, the fuel levy does not generate additional net revenues. It simply reduces the

Figure 7.1 Relationship between variable road maintenance costs and charge imposed through a fuel levy

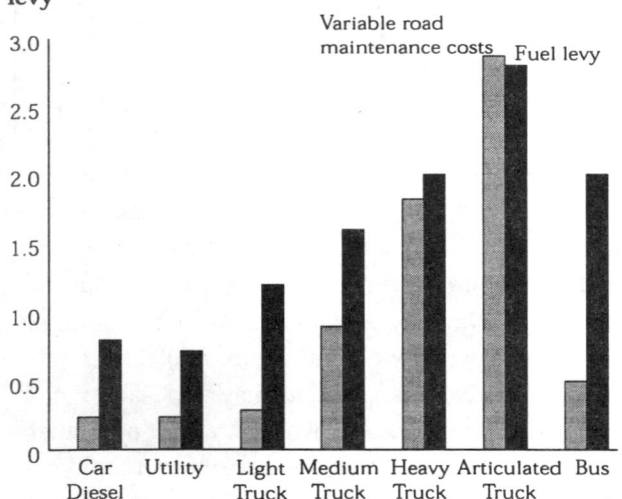

Note: Charge imposed through a fuel levy is based on a levy of $0.08 per liter (fuel consumption times fuel levy).
Source: Annex 3, table A3.4.

Figure 7.2 Fuel levies for financing roads, selected countries

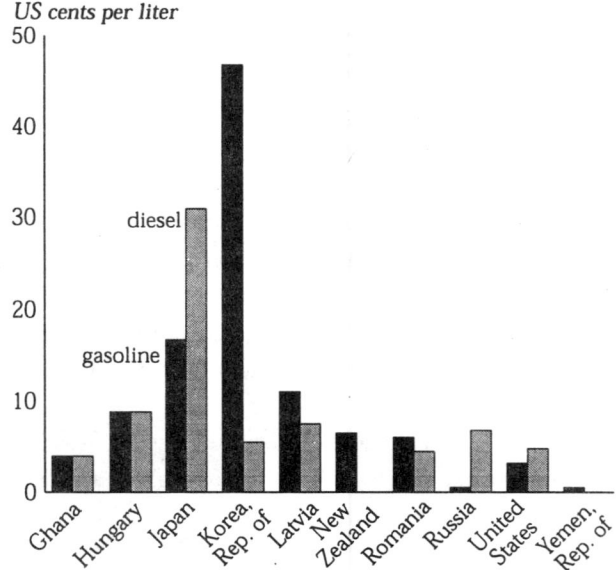

Source: Individual road funds.

Box 7.2 Collecting road-user charges under contract in New Zealand, Yemen, and Mozambique

NEW ZEALAND: All road-user charges are collected under agency agreements. The fuel excise is collected by the New Zealand Customs, which charges a fee equivalent to 0.2 percent of revenues. The sale of weight-distance charges is managed by a unit within the Land Transport Safety Authority at a cost of about 3.2 percent of revenues—just more than half is spent on collection and the balance on enforcement. The certificates are sold through the New Zealand Post, BP petrol stations, Vehicle Testing New Zealand, New Zealand Automobile Association, and AMI Insurance. Operators can also buy licenses through remote terminals. The collection of motor vehicle registration fees is managed by the Land Transport Safety Authority on a similar basis at a cost of about 18 percent of revenues.

YEMEN: The road maintenance levy, which is added to the price of fuel, is collected by the Yemen Petroleum Company. There is no formal contract, but instead a mem-orandum of understanding between the Yemen Petroleum Company and the road maintenance fund. The Yemen Petroleum Company collects the levy and deposits the revenues into the road fund on a monthly basis. Monthly reconciliation statements are sent to the road fund board showing deliveries of fuel, the levy payable, and deposits into the road fund. The Yemen Petroleum Company charges a small fee for this service.

MOZAMBIQUE: International transit fees are collected under a contract between the road fund and a freight company, The National Agency for Navigation and Freight. Transit fees are collected through the sale of bank-note-quality certificates, which are distributed by the contractor to road hauliers and then collected and endorsed at official international border crossings. The road agency supplies the certificates to the contractor, which deposits the revenue collected into the road fund bank account, less a 3 percent agency fee, within an agreed period of time.

---

implicit subsidy channeled to the road sector. To avoid subsidization, the government should ensure that pump prices are always higher than border prices.

Many OECD countries already take advantage of the low price elasticity of demand for fuel to impose higher taxes on fuel than on general consumption goods. These higher prices are justified from an economic efficiency point of view, since they generally move the overall taxation system closer to the optimum (box 7.4). Fuel prices in most developing and transition economies are generally lower than those in OECD countries. The former could therefore mobilize more domestic revenue and reduce the welfare costs of taxation by raising fuel taxes and lowering other general consumption taxes.

On the other hand, in some developing and transition economies fuel prices are already too high because tax levels are higher than the optimum or because of inefficient petroleum procurement and distribution policies. A recent survey has estimated that Sub-Saharan Africa could save about $1.4 billion a year at 1989–90 prices by rationalizing the supply of petroleum products. About half the potential savings would come from improved procurement arrangements, which would cut costs and reduce gratification payments. Another 40 percent would come from improved refining practices, while the final 10 percent would come from improved distribution and storage arrangements (Schloss 1993).

The World Bank's general advice with regard to taxes on goods and services is that they should be consolidated into a limited set of instruments with the following characteristics (World Bank 1991b):

• Revenue should be generated primarily from a broadly based tax on consumption that does not tax interindustry transactions or exports and does not differentiate by source of production (domestic or foreign).

• The best instrument to achieve this objective is a value-added tax (VAT) at a single rate of between 10 and 20 percent, with crediting provisions and zero rating for exports.

• Equity would be encouraged by introducing luxury and excise taxes with only three or four different rates on income-elastic items (such as vehicles, petroleum products, and luxuries) that are not distinguished by source of production (domestic or foreign) and by exempting items from the VAT that are a significant component of expenditures by the poor.

• Efficiency would be encouraged by additional excises and taxes on items with demonstrable negative externalities (such as "green" taxes on transport fuels).

Box 7.3 The border price of transport fuels

Border prices indicate the cheapest way to procure transport fuels. There are three main cases to be considered, countries that import refined fuel products, import crude petroleum and refine it in a domestic refinery (or pay a fee to have it refined in another country), or produce crude petroleum and refine it in a domestic refinery.

The method of calculating border prices is the same in all three cases. For countries that import refined fuel, the starting point is the f.o.b. price at the originating port, while for the other two types of countries, it is the f.o.b. price at the most efficient, available originating port (usually Bahrain, Curaçao, Rotterdam, or Singapore). The reason for choosing the most efficient, available originating port is to ensure that the costs of an inefficient local refinery, that is, inflated production costs, are not passed on to users as part of the border price, but are clearly recognized as an implicit subsidy to the refinery or local producer.

Ocean freight and port charges are then added to the f.o.b. price to produce the c.i.f. price. Alternatively, the calculation can begin directly with the c.i.f. price, which is readily available for most countries from the World Bank's quarterly report on prices of crude petroleum and petroleum products. Finally, allowance is made for inland transport costs from the port to point of sale, the distribution margin, and the sales margin. The sum of these costs represents the economic cost, or border price of fuel.

The table below shows how border prices were calculated for Brazil. The final estimates for the border prices of gasoline, diesel fuel, and kerosene were $0.23, $0.22, and $0.20 per liter, respectively, compared to actual retail prices of $0.45, $0.24, and $0.17. Diesel fuel only bears a small positive tax, while kerosene is being subsidized.

Border prices of transport fuels in Brazil, March 1989
(U.S. cents per liter)

	Gasoline	Diesel fuel	Kerosene
Price, f.o.b.	13.95	11.97	12.86
Freight charges	1.56	1.84	1.50
Insurance costs	0.47	0.58	0.54
Price, c.i.f.	15.98	14.39	14.90
Average transport costs	2.26	3.30	1.82
Distribution margin	1.87	1.30	1.28
Sales margin	3.28	3.09	2.31
Economic cost (border price)	23.39	22.08	20.31

Source: World Bank (1984).

• Tax reform will generally be more effective when accompanied by improvements in tax administration.

One of the most difficult administrative issues is ensuring that nontransport users of fuel—primarily diesel fuel—do not have to pay the fuel levy. A third or more of diesel fuel is used outside the transport sector for power generation and to operate heavy equipment in the construction, agriculture, and mining sectors. It is also used for heating. Few countries have managed to solve this problem satisfactorily. Some have applied different tax rates to automotive diesel, industrial diesel, and diesel for power generation. But it is difficult to control avoidance and evasion when several different tax rates apply to the same end product. Other countries have attempted to address the issue by offering exemptions, coloring untaxed diesel and testing road vehicles for unauthorized use, operating rebate

schemes, or compensating non-road users for having to pay the fuel levy.

Exemptions are difficult to administer, even if there are only a limited number of users, as in the power sector. Ghana used to exempt the fishing industry, but withdrew the exemption because of the high level of tax evasion. Finland, the United Kingdom, and the United States color untaxed diesel and test nonexempt diesel vehicles to ensure they are using regular (taxed) diesel.[5] Still, enforcement is difficult. New Zealand operates a rebate scheme that covers the off-road use of gasoline. The scheme is administered under contract by the Land Transport Safety Authority that, in turn, uses New Zealand Post Ltd. as its agents. Applications for refunds must be accompanied by invoices covering the purchases on which refunds are being claimed. The Land Transport Safety Authority Audit Unit undertakes ran-

Box 7.4 Strengthening revenue mobilization by improving taxation of transport fuels

Fuel prices in many developing and transition economies (such as Benin, China, Egypt, Nigeria, Romania, Venezuela, and Yemen) are currently well below those in most industrialized countries. Several countries have negligible tax rates (that is, the pump price is at or close to the border price), others simply apply standard consumption tax rates to fuel, and in others fuel prices are below the international border price (that is, the effective tax rates are negative and fuel is being subsidized).

These countries have made little systematic effort to improve domestic revenue mobilization by levying higher taxes on fuel than on other commodities. This contrasts with the practice in industrialized countries, where fuel taxes are generally much higher than general consumption taxes. A recent survey of selected OECD countries showed that gross tax rates on leaded gasoline during 1990 and 1991 were between 60 and 70 percent (75 percent in France), leading to net tax rates of 150 to 230 percent (the gross tax rate = net tax rate/[1 + net tax rate], where tax rates are measured in percent) (Creightney 1993a). This rate was three to five times higher than general consumption taxes in these countries. The available evidence on the price elasticity of demand for gasoline suggests that these differentials are justified from an economic efficiency point of view (that is, the higher rates move the taxation system closer to the optimum).

Since gross consumption tax rates are about 15 to 20 percent, gross petroleum tax rates should be between 60 and 75 percent to be economically efficient, generating net tax rates of 150 to 300 percent. In other words, if the base price of gasoline were 25 cents per liter, the fuel tax would be between 37.5 and 75 cents per liter, giving pump prices of between 62.5 and 100 cents per liter. This rate is far higher than existing gasoline taxes in most developing and transition economies. Most of these countries could therefore improve domestic revenue mobilization and reduce the welfare costs of taxation by raising fuel taxes and lowering other general consumption taxes.

dom audits to discourage fraud. In 1996 refunds amounted to about 3 percent of total revenues. Namibia is proposing to operate a similar rebate scheme.

Instead of offering exemptions or rebates, which are difficult to administer, Latvia and Mozambique compensate selected non-transport users of diesel for having to pay the diesel levy. In Latvia the Ministry of Finance estimates how much diesel the railways consumed (currently, 18 percent of total sales) and then transfers that part of the diesel levy to the railways. Likewise, farmers are entitled to receive annual compensation equivalent to 120 liters of diesel fuel for every hectare of land under cultivation (120 liters being the estimated amount of diesel used to cultivate one hectare of land). The local municipality assesses the applicable land area. A similar compensation scheme applies to the fishing industry. Mozambique uses an even simpler method to compensate farmers. Twenty percent of the diesel levy is paid into a special fund, which provides financial support for agriculture. There are thus a number of ways to ensure that the fuel levy is paid only by road users or that they are compensated for having to pay it. The key issue is to decide which method is likely to work best in each country context.

The final administrative concern relates to relative fuel price distortions. Fuel levies raise fuel prices, which may encourage substitution between different transport fuels. The biggest problem arises with kerosene. Some governments keep kerosene prices low to minimize the impact on low-income households that use it for cooking and lighting. They also keep kerosene prices low to encourage substitution of kerosene for fuelwood to reduce deforestation. Kerosene can be mixed with either gasoline or diesel fuel and, when mixed with a little engine oil, can even be used as a complete substitute for diesel fuel. A high price differential between diesel and kerosene will thus encourage substitution, and the fuel levy will not realize its full potential. The only ways to discourage substitution are to color kerosene and inspect vehicles for mixing or to issue coupons to poor households to purchase kerosene at concessionary rates. Neither solution is entirely satisfactory. The best option is to avoid wide price differentials between kerosene and diesel fuel.

Fuel Smuggling

Financing roads through fuel levies breaks down when there is rampant fuel smuggling—a major problem in North America, North Africa and the Middle East, parts of Asia, and Africa, where low fuel prices in some countries have led to massive fuel smuggling. Canadians buy cheap fuel in the United States, Algeria smuggles fuel into Tunisia, Yemen smuggles fuel into Saudi Arabia, some provinces in China smuggle fuel into other provinces, and Nigeria smuggles fuel into most West African countries. Indeed, in 1992 it was estimated that between one-quarter and one-half of the fuel consumed in Cameroon and Benin was smuggled from Nigeria. Smuggling makes it virtually impossible for governments to mobilize any revenues by taxing imported fuels.

There is no easy way around this problem. Wide disparities in prices lead to large potential profits and hence to widespread bribery and corruption. Attempts to prevent smuggling cannot rely on enforcement alone. Three alternatives currently being tried include: removing subsidies and exchange rate distortions, making the currency nonconvertible so that the sale of smuggled fuel is more difficult, and introducing network-wide road tolls in lieu of the fuel levy. Algeria has removed the bulk of its fuel subsidies, which has reduced large-scale smuggling to a trickle. In Yemen the realignment of the official exchange rate, combined with an increase in fuel prices, has also reduced fuel smuggling. Convertibility has been suspended in Cameroon and Benin, and Cameroon has introduced road tolls over the entire main road network. However, the tolls were not intended primarily to discourage smuggling, and it is estimated that 75 percent of potential toll revenue is lost through evasion and leakage. Unless toll systems are carefully designed and administered in collaboration with the road transport industry, they will face public hostility and are unlikely to generate much revenue.

Pricing and Cost Recovery Policies

This section focuses on ways of recovering the costs of maintaining, improving, and rehabilitating the road network and on ways of using congestion charges to ration scarce road space. It does not deal with the costs of other externalities—like noise, ground water pollution from de-icing chemicals, and greenhouse gasses— since the government should handle such externalities directly through regulations and corrective taxes. (These should take the form of an additional environmental levy added to the price of transport fuels.

Finland, Norway, and Sweden all add various environmental taxes to the price of gasoline and diesel to discourage lead in petrol, encourage sulfur-free diesel fuel, and reduce CO_2 emissions.) The pricing and cost-recovery policies discussed in this chapter have four objectives: to provide the correct market signals to road users, to ensure that road agencies use resources efficiently, to constrain the size and quality of the road network to what is affordable, and to generate sufficient revenues to operate and maintain the core road network on a sustainable long-term basis. These policies must therefore balance several conflicting objectives.

Basic Principles

To maximize net economic benefits, road-user charges should be set equal to the costs of the resources consumed when the road network is used. These costs are generally referred to as short-run marginal costs. Two costs must be considered: the cost of damage done to the road surface by the passage of vehicles (that is, the variable costs of operating and maintaining the road network) and the additional costs that each road user imposes on other road users and on the rest of society— primarily the costs of road congestion. Congestion is the classic negative externality in the road sector and is the one normally taken into account when estimating the optimal user charge.[6]

The basic principle behind efficiency pricing is that additional road capacity should be financed through congestion charges. Capacity should be expanded when the annual costs of road congestion are equal to the annualized costs of expanding capacity. But since less than half the costs of operating and maintaining the road network vary with traffic (see table 7.3)—and roads in developing and transition economies do not generally experience widespread and persistent road congestion outside large urban areas—if prices are set equal to short-run marginal costs, large financial deficits will ensue. Furthermore, since most governments in developing and transition economies are acutely short of fiscal revenues, it is rarely possible for them to finance these deficits through general taxation. The funds are simply not available.

How should these deficits be financed? The obvious targets are road users and, in the case of local access roads, those who benefit from road access. Furthermore, given the relatively high welfare costs of mobilizing general tax revenues and the low price elasticity of demand for road use, there is a prima facie case for assuming that the welfare costs of raising most of the required revenues from road users are lower than the costs of mobilizing them through general taxation. There are also distributional arguments in favor of raising most of these revenues from road users. Road users are among the wealthiest members of society. Although the poor depend heavily on public transport for job searching and gaining access to public services, it is better to assist them by subsidizing selected transport services or by providing other forms of income support.

Attempting to achieve full cost recovery is consistent with the desire to link revenues and expenditures to subject the road agency to a hard budget constraint. If some costs are financed through subsidies or other transfer payments, market discipline is weakened. Pressure to keep costs under control—and only undertake expenditures for which users are willing to pay—requires a clear market signal that forces road users to recognize the full costs of providing road services. The road tariff should therefore reflect the costs of operating and maintaining the road network, and increased road spending should automatically raise the road tariff, even though it will usually also reduce VOCs. Imposition of a hard budget constraint thus requires full cost recovery from road users and, in the case of local access roads, both road users and the beneficiaries of road access.

This leads to three basic pricing and cost-recovery policies:
• Never set the road tariff—ideally, the variable element of the road tariff—lower than the variable cost of operating and maintaining the road network.
• Ensure that the road tariff and the taxes and charges used to support local access roads collectively cover all road costs.
• When there is significant road congestion, the road tariff should also include congestion costs. (This rule will generally apply only to a handful of seriously congested cities.)

Practical Considerations

There are three main practical problems. First, the variable costs of maintaining different types of roads differ significantly—from about 0.026 to 0.177 cents per

Table 7.3 Costs of road maintenance on different types of roads with typical loading conditions
(U.S. cents per vehicle per km)

	Main roads		Local access roads		
	Major arterial, paved	Minor arterial, paved	Low volume, paved	High volume, unpaved	Low volume, unpaved
Traffic (ADT)	10,000	3,000	300	300	50
Deflection (mm)	0.5	1.0	1.5	—	—
Medium loading, high motorization (10 percent trucks)					
Variable costs	0.026	0.110	—	—	—
Fixed costs	0.198	0.548	—	—	—
Total	0.224	0.658	—	—	—
Medium loading, average motorization (50 percent trucks)					
Variable costs	—	0.177	1.449	—	—
Fixed costs	—	0.635	2.850	—	—
Total	—	0.812	4.299	—	—
Primary gravel, average motorization					
Variable costs	—	—	—	1.035	1.321
Fixed costs	—	—	—	1.058	6.351
Total	—	—	—	2.093	7.673

— Atypical loading conditions that have not been calculated.
Note: Based on 49 data sets from 33 developing and transition economies, with medium vehicle operating costs, medium road agency costs, and temperate environmental conditions. Medium loading means that the average design loading in the 33 countries included in the sample. U.S. cents per vehicle per km is the average cost for all vehicles.
Source: Calculated in annex 3.

vehicle-km on the main road network, to 1.499 cents per vehicle-km on high-volume local access roads, and from 1.035 cents to 1.321 cents per vehicle-km on the rural road network (table 7.3). Total costs likewise vary from a low of 0.224 cents per vehicle-km on the main network to a high of 7.637 cents per vehicle-km on the rural road network. Charges based strictly on costs would thus involve wide differentials between different types of roads and different central and local government road agencies. This is simply not practical, although it is possible to maintain some differential between urban and rural areas and among different regions. A practical set of user charges will thus involve a great deal of averaging.[7]

Further, the variable costs of maintaining the road network also differ significantly with vehicle type (see figure 7.1). Cars impose relatively small costs on the road network, while articulated trucks impose costs twelve times larger. In principle, an articulated truck should therefore pay 12 times more than a car. But if the main charging instrument is a fuel levy it will only pay three-and-a-half times as much (an articulated truck uses about three-and-a-half times as much fuel as a diesel car). The available charging instruments therefore introduce further averaging that can be avoided only by switching to weight-distance fees, which can be more accurately calibrated to reflect underlying road-use costs.

The final practical problem relates to the way license fees and the fuel levy are set to ensure that to the extent feasible, the variable element of the road tariff paid by each class of vehicle (the fuel levy) covers the variable costs that vehicle imposes on the road network, and the road tariff and the taxes and charges used to support local access roads collectively cover all road costs. The fuel levy, by itself, will generally undercharge articulated trucks and overcharge other vehicles, particularly buses. The license fee must therefore be used to compensate. In other words, the license fee cannot be used strictly as an access fee set to cover only fixed costs. The available pricing instruments are too blunt for that. Instead, the combined license fee and fuel levy have to be set to ensure that each vehicle class covers the variable costs it imposes on the road network. This then results, not in a strict two-part tariff, but in a quasi-two-part tariff. Clearly, with these charging instruments there

is no scope for using the inverse elasticity rule (that is, Ramsey pricing), although it would be applicable for weight-distance fees. (See annex 2 for a description of the inverse elasticity rule. Annex 3 provides an example illustrating how to estimate the above two-part tariff.)

Financing Maintenance

The above model suggests that the costs of operating and maintaining the interurban road network should be financed through the road tariff, that in urban and rural areas at least the variable costs of operating and maintaining the road network should be financed through the road tariff, and that the balance of the required expenditures in urban and rural areas should be financed from local revenues. These local revenues may come from parking charges, particularly in large urban areas (these are usually a minor source of revenues); local property taxes (a major source of revenues since road access increases the tax assessment value of the property); market and/or product taxes (effectively a local sales tax); and other miscellaneous taxes. In rural areas and informal urban settlements the local community sometimes contributes materials and/or volunteer labor in lieu of such taxes.

One of the key features of the above financing arrangement is that it focuses attention on the affordability of a fully funded road maintenance program and hence on the need to define a core road network that users are willing and able to fully finance. Most countries are now having to face this issue. Road networks—particularly in Africa, Asia, and Latin America—expanded too rapidly during the 1960s and 1970s, and governments can no longer afford to maintain them in full. Instead, governments are being forced to define an affordable core network. Noncore roads either receive minimal maintenance or are handed over to lower levels of government. A typical core road strategy involves fully maintaining all major roads in good or fair condition and carrying out only emergency and spot maintenance on roads in poor condition.

Financing New Investment

New investments include making road improvements (such as paving a gravel road), extending the road network (such as constructing an agricultural penetration track), and expanding road capacity (such as widening a road). There are sound economic reasons for wanting to finance improvement and extension by taxing those who benefit. There are also sound economic reasons for wanting to finance increased road capacity on congested roads through congestion charges (see the section above).

In the case of interurban roads the bluntness of the available charging instruments makes it difficult to confine charges to beneficiaries, except on roads carrying high volumes of traffic and that lend themselves to tolling (urban road congestion is dealt with below). The choice of financing instruments for the overall interurban network thus boils down to either financing all new investments from general taxes channeled through the government's development budget or financing new investments by charging all road users.

People have strong views on which option to use. Many believe the road tariff should finance only operation and maintenance of the road network and that all new investment should be financed through the government's development budget. Their concern is that that new construction might otherwise take precedence over maintenance or that the road agency might overinvest. There is some evidence to support this view—many countries continued to build new roads during the 1980s at the expense of maintenance.

An additional concern is that major new investments in the interurban road network generally have substantial effects on land-use patterns, the location of industry, and adjoining property values. Since this issue raises strategic and political concerns, such investment decisions should be made and financed by the government. On the other hand, there are also arguments in favor of financing new investment by charging road users. Only by forcing road users to pay the full costs of using the road network—including the costs of investment—will the size of the network be constrained to what is affordable and will essential investments be carried out regardless of the state of the government's budget.

There is no simple answer. Some countries finance most new investments through the development budget (including Guatemala, Malawi, Yemen, and Zambia), while others finance some investment through user charges (including Georgia, Hungary, Japan, Korea, Latvia, New Zealand, Romania, Russia, and South

Africa). It is really a question of governance. In countries where new investments are frequently made for political reasons and where the roads board (if any) is unable to stand up to these political pressures, new investments should probably be financed through the development budget. Where there are strong, representative road boards that are able to withstand political pressure, it may be better to finance new investments through the road tariff. This will ensure that all new investments are subjected to the test of the marketplace. A strong representative board should also be able to ensure that new investment does not displace maintenance.

Slightly different considerations apply for local government roads. The overriding objective concerning new investment is to ensure that local governments undertake only priority projects, not those for which funds are provided as a grant channeled through the government's development budget. This issue argues in favor of a matching-grant system. Local governments should have to demonstrate the priority of their investment programs by paying part of the costs from local revenues. These revenues can come from land-value increment taxes (that is, betterment taxes and frontage levies), charges to adjoining landowners, urban congestion charges (where applicable), or other forms of property tax. The balance of the expenditures are then financed by the road tariff or through the government's development budget. The amount financed by the local government should clearly be based on ability to pay.

Financing Road Rehabilitation

Most countries have large backlogs of deferred maintenance. Further, governments are short of fiscal revenues and are generally unable to finance much road rehabilitation from their own resources. So, where will the funds come from? First we must recognize that most developing and transition economies cannot afford to rehabilitate all roads that are in poor condition. The best they can hope for is to rehabilitate a core network that the country can afford to maintain on a sustainable long-term basis. The remaining roads will either have to receive minimal maintenance or be handed over to lower levels of government and local communities. But even rehabilitating core roads will cost several billion dollars each year for the foreseeable future. There are three possible means of financing:

reallocating existing spending from new construction to rehabilitation, seeking donor-financed loans and grants, and relying on the road tariff.

The first option offers little hope. Few developing and transition economies have oversized construction programs. Most new construction is being undertaken in rapidly growing economies that are suffering from serious road congestion (China, India, Korea), or in transition economies with outdated road networks in urgent need of modernization (Russia). So, there is limited scope for reallocating domestic resources from construction to rehabilitation. The second option, donor financing, is already being used, with donors currently financing at least one-half of the required rehabilitation programs.

This money is, however, not free. True, some comes in the form of grants and some comes in the form of concessionary loans, but governments still have to service these loans. In the short term most governments are doing so from general tax revenues. In other words, other sectors are being taxed to finance road rehabilitation programs. This practice is not sustainable under present fiscal conditions and in the long term. And donor financing will not be available indefinitely. Thus there is only one realistic long-term option: financing road rehabilitation programs through the road tariff.

There are three qualifications. First, funds for rehabilitation must be clearly designated as a surcharge in the road tariff, and the surcharge should eventually be phased out. Second, the costs can be spread and made more affordable by continuing to use international and domestic borrowing. And third, the required funds should be increased gradually to enable local consultants and contractors to build up their capacity. The decision to borrow should nevertheless be based on a careful assessment of alternative financing options and their costs.

Managing Urban Road Congestion

Congestion charges can be used to manage urban traffic and generate additional revenues for investment. The simplest way to start charging for congestion is through parking charges, supplemented by improved traffic management to prevent the parking charges from spilling over into illegal parking and other avoid-

ance strategies. Parking charges offer a natural transition from the use of physical measures to improve road capacity to the use of congestion charges to ration scarce road space.

Once the value of parking charges has been exhausted, serious consideration should be given to introducing an explicit road pricing scheme, although the only serious attempts made so far to develop road pricing instruments are in Hong Kong, Norway, Singapore, and Sweden (see box 7.5). Much of the public resistance to these road pricing schemes centers on how the revenues are used. Norway and Sweden have largely overcome this resistance by dedicating the revenues to improving urban transport services.

The Likely Structure of User Charges

A sustainable road maintenance program generally requires vehicle license fees, which vary from about $150 per year for a car, to $500 to $600 per year for a bus or medium truck, to about $2,000 per year for an articulated truck. These license fees must be combined with a fuel levy of about $0.05 to $0.10 per liter to ensure that the costs of operating and maintaining the road network can be fully funded. License fees in most countries, particularly those applicable to heavy vehicles, are generally lower and should be raised and/or supplemented by a heavy vehicle license fee to generate the required revenues.

The same is not true of fuel levies (see figure 7.2). A number of countries either have, or are well on their way to having, fuel levies of $0.10 per liter, while some have fuel levies well in excess of this rate (Japan and Korea). Furthermore, a fuel levy of $0.10 per liter does not necessarily make fuel unduly expensive. Most developing and transition economies have diesel prices of about $0.55 per liter or less and gasoline prices of about $0.70 per liter or less (figure 7.3). A $0.10 fuel levy would thus still leave fuel prices in most developing and transition economies at levels lower than those in Europe and Japan. They would also be lower than the prices based on the EU's recommended minimum excise duty rates on motor fuel (1998: gasoline $0.473 per liter, diesel $0.344 per liter; 2000: gasoline $0.500 per liter, diesel $0.381 per liter; 2002: gasoline $0.555 per liter, diesel $0.436 per liter). Only in a few coun-

---

Box 7.5 Methods of charging for urban road congestion

There are four main ways of using pricing to reduce urban road congestion: charging for parking, imposing a higher license fee or fuel levy on urban road users, charging a fee for entering the urban road network, or charging for the use of individual streets or designated parts of the urban road network. This box describes the last two methods.

Entry fee systems charge vehicles each time they cross a cordon. Fees can be collected manually or electronically. The schemes either use tollgates to charge vehicles entering the restricted zone, as in Bergen, Oslo, and Trondheim in Norway, or use area licenses, as in Singapore. With area licenses, there is no need for tollgates. Vehicles simply display a supplementary prepaid license when entering and operating within the restricted zone. General road pricing, in which vehicles are charged either on individual routes or when using parts of the road network, is feasible only with electronic charging schemes such as: automatic vehicle identification (AVI), electronic number plate (ENP), and Smartcard. The vehicles equipped with an AVI tag, an ENP, or a Smartcard are identified when they pass an electronic reader. The reader charges either the vehicle's account (precredited or not) or the prepaid Smartcard itself. Oslo and Trondheim in Norway use both manual and electronic tolling systems. Users can thus choose either to subscribe to AVI and be identified or use the manual toll lanes and remain anonymous. In Trondheim 95 percent of users use the electronic tags. The charging rates differentiate between peak and off-peak hours. Revenues are used to finance improved transport infrastructure, including the transit system and pedestrian and bicycle facilities. The ENP scheme has been tested in Hong Kong, and the Smartcard system is set to be introduced in Singapore in 1998. Electronic charging schemes do away with the need for toll plazas and reduce delays.

On the basis of benefit-cost calculations the labor-intensive technology of supplementary licensing outranks the capital-intensive technology of electronic pricing through AVI and is especially suitable for those developing and transition economies that have a large pool of unemployed workers and a limited number of urban access routes. Cordon pricing through standard manual tollgates and unattended reserved lanes (as in Bergen) may prove to be a worthwhile, labor-intensive technology for developing and transition economies. Electronic pricing through AVI is a viable alternative for newly industrializing economies whose standard of living is rising, but where rapid urbanization and growth of motorization pose major problems. Smartcard technology is not widely available on a commercial basis and is not yet recommended for developing and transition economies—even though Singapore is currently testing an Electronic Road Pricing system with smartcards and will change to this system in 1998 if the trials prove successful.

Source: Hau (1992).

---

tries might the introduction of a high fuel levy need to be accompanied by revision of the underlying fuel tax structure to ensure that the final price of fuel was not unreasonably high.

Key Recommendations and Conclusions

The key elements of the strategy required to generate a secure and stable flow of funds for roads include the following:

• Choose pricing instruments that send a clear market signal to road users. The signals should be easily recognizable, related to road use, easy to separate from indirect taxes and service fees, and simple to administer.

• A review of the various charging instruments available suggests that road users should be charged for use of the road network by way of a two-part tariff. Vehicle license fees can be used to charge for access to the road network, while fuel levies, international transit fees, and tolls can be used to charge for use of the road network. Fuel consumption is not exactly related to variable road maintenance costs, but it is related closely enough for practical charging purposes.

• To improve the effectiveness of road-user charging instruments and to enhance revenue mobilization. it is desirable to simplify the fee structure (simplify administration and reduce the costs of collection and compliance) and to collect more fees under contract with the private sector.

• To ensure there are no inadvertent subsidies, all vehicles should be required to pay the two-part tariff, and the government should reimburse the road agency for mandated exemptions (such as for diplomatic vehicles). The pump price of fuel should also be set above the applicable border price of fuel.

Figure 7.3 Fuel prices in selected countries, early 1997

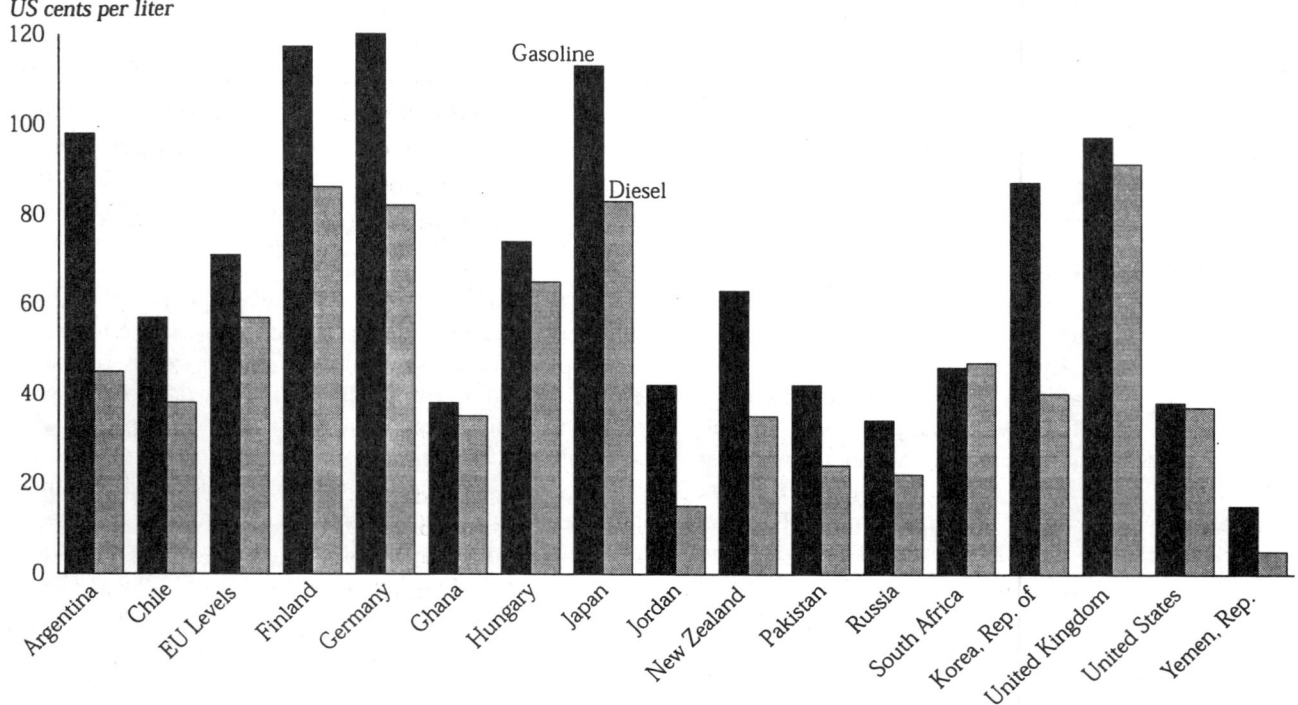

Note: Recommended 1988 EU levels based on border prices of $0.24 per liter for gasoline and $0.23 per liter for diesel.

• Countries should consider taking advantage of the low price elasticity of demand for fuel by imposing higher taxes on fuel than on general consumption goods. Doing so would move the overall taxation system closer to the optimum, particularly in countries where fuel prices are well below those in OECD countries. Such countries could improve domestic revenue mobilization and reduce the welfare costs of taxation by raising fuel taxes and lowering general consumption taxes.

• Since a third or more of diesel is used outside the transport sector, it will probably be necessary to ensure that non-transport users do not have to pay the fuel levy. Applying varying tax rates to different end users, offering exemptions, coloring untaxed diesel, and operating rebate schemes are all difficult to administer. Compensating road users for having to pay the fuel levy, as applied in Latvia and Mozambique, is one of the easiest methods to administer. The key issue is to decide which method is likely to work best in each country.

• To discourage the mixing of kerosene (which is not subject to the fuel levy) with gasoline and diesel, wide price differentials between kerosene and diesel should be avoided if possible.

• Since it is virtually impossible to finance roads through fuel levies when fuel smuggling is widespread, steps should be taken to discourage smuggling. There is no easy answer. Of the options available, removing subsidies and exchange rate distortions are among the most effective. Using road tolls as an alternative way of raising road revenues is the least effective.

• The two-part tariff should be set to: provide the correct market signal to users, ensure road agencies use resources efficiently, constrain the size and quality of the road network to what is affordable, and generate sufficient revenues to operate and maintain the core network on a sustainable long-term basis.

• To maximize net economic benefits, road-user charges should be set equal to the costs of the resources consumed when using the road network. These costs consist of the variable costs of operating and maintaining the road network and the costs of road congestion (the costs of other externalities should generally be handled through regulations and corrective taxes, which should take the form of an additional "green" environmental levy added to the price of transport fuels). But since less than half the costs of operating and maintaining the road network vary with traffic, prices set equal to these costs would generate large financial deficits. There are strong arguments in favor of financing these deficits from charges to road users and those who benefit from road access.

• There are three basic principles for setting up a two-part tariff: never set the road tariff—ideally, the variable element of the road tariff—lower than the variable costs of operating and maintaining the road network; ensure that the road tariff and the taxes and charges used to support local access roads collectively cover all road costs; and if road congestion is significant, the road tariff should also include congestion costs (although this will generally apply only to a handful of seriously congested cities).

• The road tariff will require a great deal of averaging among different road types and different vehicle types. The fuel levy will generally undercharge articulated trucks and overcharge other vehicles, particularly buses. The license fee must therefore be used to compensate for this and cannot be strictly used as an access fee set to cover only fixed costs. The available pricing instruments are too blunt for that. What results, then, is not a strict two-part tariff, but a quasi two-part tariff.

• The revenues from the above road tariff are normally used to finance the entire cost of operating and maintaining the trunk road network and, on a cost-share basis, part of the cost of local government roads. The costs of new investment are either financed through the government's development budget or, in countries with good governance, through the road tariff. Investment in local government roads again tends to be financed on a cost-share basis to provide the correct incentives to local governments. A great deal of road rehabilitation is currently financed through donor loans and grants. In the longer term it will have to be financed through the road tariff, using international and domestic borrowing to spread the burden over several years.

• Parking charges are the simplest way to charge for congestion. Once the value of parking charges has been exhausted, serious consideration should be given to introducing an explicit road pricing scheme, bearing in mind that much of the public resistance to such pricing schemes is related to what is done with the revenues. Revenues should ideally be dedicated to improving urban transport services.

• The above pricing and cost recovery policies are likely to result in vehicle license fees that vary from about $150 per year for a car to $500–$600 per year for a bus or medium truck, to about $2,000 per year for an articulated truck. These fees must be combined with a fuel levy of about $0.05 to $0.10 per liter to ensure that the costs of operating and maintaining the road network are fully funded through the road tariff.

Notes

1. In developing and transition economies, where virtually all governments are critically short of fiscal revenues, improved cost recovery is generally more important than improved demand management.

2. This emphasis counteracts the standard presumption of economic theory that public sector production is efficient and that costs, including marginal costs, are minimized (see Kranton 1990). Although the inverse elasticity rule remains valid when costs are not minimized, welfare is no longer maximized by setting the ratio of price over marginal cost proportional to the sum of the inverses of the supply and demand elasticities. Reducing marginal costs to increase welfare also becomes important.

3. Service fees cover the costs of establishing title to property (to facilitate law enforcement), checking vehicles for mechanical soundness, and monitoring payment of license fees. As such, they are not user charges and should be set only to cover servicing costs.

4. Experience with parking charges is not encouraging. For example, in 1991 the Nairobi City Council earned $17,500 from car parks and parking meters, but it cost $82,000 to operate and maintain these facilities.

5. In the United States some vehicles, like school buses, are exempted from paying the fuel levy. Also, off-road use by industry and agriculture is exempted.

6. External disbenefits also include the road damage externality. Each vehicle damages the road pavement, which increases the VOCs of all subsequent vehicles that use the road. But if the road network has a fairly uniform age distribution, and if maintenance policies are condition-responsive, road damage externalities are zero when traffic growth is zero and all road damage is caused by vehicles. Road damage externalities are negligible in all other reasonable cases (see Newbery and others 1988).

7. This is true in many sectors. In the case of electricity the costs of generating the base load are estimated by pooling the costs of individual power stations and calculating the average variable and fixed costs for the entire group. The variable costs of a hydropower station (which are virtually zero) are thus pooled with those of coal, oil, and gas-fired stations, and with stations of different ages.

8. Managing the Revenues

We start with the assumption that the revenues allocated for roads will be separated from the government's consolidated budget and managed on a stand-alone basis. This requires establishing a road fund—and some kind of road fund administration—to actively manage the revenues. This chapter focuses on the desirability of linking road-user revenues with road expenditures, the different types of road funds, the characteristics of existing road funds, the typical problems affecting conventional (first-generation) road funds, and how to set up a commercially managed road fund.

Linking Revenues and Expenditures

Nearly all road funds established during the 1970s and 1980s were set up during periods of fiscal stress and were designed to deal with failed budgetary systems. These were called the "first-generation" road funds. They relied primarily on earmarked revenues—which often included general taxes as well as road-user charges—and were set up to protect the road sector from the vagaries of the government's budgetary process. The ultimate objective was to ensure that road maintenance could be adequately funded.

The road funds that were restructured or set up during the 1990s had a different rationale. They were introduced primarily as part of a long-term strategy designed to commercialize the road sector. The idea was to bring roads into the marketplace, put them on a fee-for-service basis, and manage them like a business (box 8.1). Road users would then pay for using roads, and the revenues collected would be slated to finance road development and maintenance. Indeed, pricing and cost-recovery policies will influence demand and strengthen market discipline only when this link is made. Spending on

roads then becomes dependent on users' willingness to pay, which helps to impose a hard, but fair, budget constraint on the agencies supplying road services.

Road users are generally willing to pay for roads only when the money is actually spent on roads and the work is done efficiently.[1] To that end, it is important to recognize that the money paid into the road fund should not include any earmarked tax revenues. Road fund revenues should consist only of charges for use of the road network: vehicle license fees, supplementary heavy vehicle license fees, international transit fees, bridge and ferry tolls, fines for overloading (or at least the part of the fine that represents damage to the road pavement), and fuel levies. These charges would make up the road tariff—showing that roads are being treated like any other public enterprise—and should not be confused with the general taxes that road users also have to pay and that should continue to be paid into the government's consolidated fund. This is one of the main reasons why putting roads on a fee-for-service basis—by introducing a road tariff and depositing the proceeds into a special account—is not the same as conventional earmarking.

Types of Road Funds

Road funds come in all shapes and sizes (tables 8.1 and 8.2). Their objective is nearly always to provide regular finance to support spending on roads (often confined to maintenance), keep the revenues separate from the government's consolidated account, and account for use of these funds. Some road funds finance only national or main roads (South Africa); some finance only state, provincial, and regional roads (Argentina, the state road and transportation funds in

Box 8.1 Conventional earmarking versus commercialization

Earmarking is the practice of setting aside revenues raised from certain taxes to cover specified public expenditures. Many economists argue that earmarking imposes undesirable rigidity on government spending decisions and should be discouraged. For example, it is inefficient to set aside, say, 20 percent of overall fuel tax revenues to finance national roads, since not all fuel consumption is related to road use. The required expenditures will generally be larger or smaller than this amount, only part of the revenues will come from taxes related to road use, and it may be desirable to use some or all of the fuel tax revenues for other purposes.

But others argue that earmarking taxes under certain circumstances can improve allocative efficiency in that they act as surrogate prices when the taxes chosen are levied only on those who benefit from the expenditures. For example, in both the United States and Japan part of the gasoline tax and other motor vehicle tax proceeds are earmarked for the road fund, the income from which is used to meet the costs of operating, maintaining, improving, and extending designated parts of the road network. Many argue that such earmarking is a helpful device for approximating benefit taxation and will promote more efficient expenditure decisions.

Nevertheless, this paper does not propose the above type of earmarking. Instead, it favors commercializing the road sector by managing it along lines that mirror the management of comparable enterprises in the private sector. In this context the road-user charging system should derive its revenues only from road-user charges (not from a proportion of any sales or excise taxes), extract no revenues from other sectors, adjust charges regularly to meet anticipated expenditure requirements, keep the revenues apart from the consolidated fund, and use the revenues only to finance the road services paid for by the road users. This institutional distinction has important implications for efficiency. The charges create a constituency for the agency supplying the service (that is, they create a specific, albeit surrogate, market), make the agency more accountable to its users, and, by clearly linking revenues and expenditures, help to impose a hard budget constraint on the road agency. Specifically, the proposed financing arrangements differ in that the road tariff is:

- Set to achieve specific objectives, including demand management and cost recovery for particular services.
- Not set in relation to the government's overall fiscal targets (although usually collected under the government's tax-making powers).
- Added to pre-existing standard sales and excise taxes or, when fuel is highly taxed, partly replace and partly add to pre-existing taxes.
- Used to impose a hard budget constraint on the agency supplying road services.

The road tariff is generally set by a public-private management board that recommends the charges to the ministry of finance, the charges are set to ensure that each vehicle cover the costs it imposes on the road network, and all vehicles collectively cover the entire costs of operating and maintaining the road network.

---

the United States, the Russian regional road funds, and the regional municipal road funds in Latvia); and some have been set up as urban road funds to finance only urban roads (box 8.2). Most, however, finance expenditures on the whole road network.

Some road funds also finance nonroad expenditures. The Korean traffic facility special account (the expanded 1994 version of the 1989 road sector special account) includes a special account for urban rail, express rail, airports, and harbors, while the U.S. Federal Highway Trust Fund finances community road safety programs, high-speed rail lines, and bike trails and makes transfers into a mass transit fund. The Latvian state road fund finances passenger bus subsidies, while the New Zealand road fund finances passenger transport, the Land Transport Safety Authority, and police enforcment of road safety, before the balance is transferred to Transfund. Furthermore, although Transfund primarily finances roads, since 1996 it has also started to finance local authority "alternatives to roading."

Both Korea and South Africa have unusual road funds. The Korean road fund, unlike others, is not really a separate account, but a mechanism for channeling earmarked revenues to finance the regular road budget. The funds are not managed separately from the consolidated fund, and there are no special financial procedures or separate auditing arrangements. The South African road fund, on the other hand, was financed through a fuel levy until 1988, when the levy was suspended. Since then it has operated as a mechanism to manage finance for the national highway network, which now comes in the form of an annual grant from the consolidated fund. A road fund can thus operate simply as a means to increase transparency and strengthen financial discipline. It does not have to be tied to dedicated financing. The fuel

Table 8.1 Legal and administrative arrangements applicable to selected road funds

Country	Legal basis	Oversight	Type of entity	Own staff	What does it finance	Main source of revenues
Ghana	Decree 1985, legislation 1996	Public - private board	Separate agency	Yes	All expenditures	Fuel levy, transit fees, vehicle fees
Guatemala	Legislation 1993	Public - private board	Separate agency	Yes	Maintenance of national roads only	Fuel taxes, vehicle fees, tolls, miscellaneous
Hungary	Cabinet decree 1989, state law 1992	Road agency	Division of road agency	Yes	All expenditures on state roads plus transfers to municipalities	Fuel levy, weight-related vehicle tax, donor finance
Japan[a]	Special account law 1954	Road council	Division in Road Bureau	Yes	All expenditures on national roads plus transfers to local governments	Gasoline tax, liquid petroleum gas tax, vehicle tonnage tax, general budget
Korea, Rep. of[b]	Special account law 1989, amended 1994	Ministry of Construction and Transportation	n.a.	No	All expenditures on national roads, some expenditures on expressways and provincial roads	Fuel tax, excise tax, tolls, general budget
Latvia[a]	Cabinet decree 1994	Public-private advisory board	Division of road agency	Yes	All expenditures on state roads plus transfer to municipalities	Fuel tax, vehicle fees, general budget
New Zealand	Legislation 1953, amended 1996	Primarily private board	Separate agency	Yes	All expenditures	Weight-distance charges, fuel levy, vehicle fees
Malawi	Legislation 1997	Public - private board	Separate agency	Yes	All expenditures, maintenance priority	Fuel levy, vehicle licenses, transit fees, overload fines
Romania	Legislation 1996	Ministry of Transport	Division in Ministry of Transport	Yes	All expenditures plus transfers to counties and villages	Fuel levy, vehicle sales tax
Russia[a]	Legislation 1992	Federal Highway Department	Division in Highway Department	Yes	All road expenditures plus transfers to regions	Fuel and lubricant tax, vehicle sales tax
South Africa	Legislation 1935, plus amendments	Public - private board	Staff in director's office	Yes	All expenditures on national roads	General budget since 1986
United States[a]	Legislation 1956	Committees of Congress	Accounting mechanism managed by Treasury	Yes	Primarily capital works on federal-aided highways	Fuel tax, vehicle sales tax, heavy-vehicle tax
Yemen	Presidential decree 1995, ratified by Parliament	Civil service board[c]	Separate agency	Yes	Maintenance only	Gasoline levy, overload fines, general budget

n.a. Not applicable.

Note: In addition to the road funds shown in this table, there are national road funds and earmarking devices in Belgium, Luxembourg, the Netherlands, Switzerland, Benin, Central African Republic, Chad, Kenya, Madagascar, Mozambique, Rwanda, Senegal, Sierra Leone, Tanzania, Zambia, Zimbabwe, Argentina, Guyana, Honduras, Azerbaijan, Czech Republic, Georgia, Kazakhstan, Lithuania, Mongolia, Slovak Republic, Turkmenistan, Ukraine, and Uzbekistan. New road funds have recently been set up in Jordan and Namibia, and Armenia is currently establishing one. The Swedish government is also considering whether the Swedish National Road Administration should in the future be financed solely through vehicle license fees and a fuel levy of about $0.08 per liter.

a. National or federal road fund.

b. Road fund is not a separate account but a mechanism for financing the regular road budget.

c. Includes nonvoting members from the Chamber of Commerce and Ministry of Transport.

Table 8.2 Financial arrangements applicable to selected road funds

Country	Fuel levy	Annual revenues[a] (million dollars)	Adjusting charges	Deposit mechanism	Financial procedures, regulations	Auditing
Ghana	$0.05 per liter	60 (1997)	By board	Direct deposit	Yes	Auditor General or independent audit
Guatemala	$0.023 per liter	32 (1997)	By annual budget	Through consolidated fund	Yes	Auditor General
Hungary	$0.095 per liter	233 (1997)	By annual budget	Through consolidated fund	Yes[b]	Auditor General
Japan[c]	25 percent of gas tax	30,000 (1995)	By tax law every 5 years	Road fund is a line of credit	Yes	Independent audit
Korea, Rep. of[d]	67 percent of gas and diesel tax	5,000 (1996)	By tax law	n.a.	n.a.	Auditor General
Latvia[c]	50 percent of gas and diesel tax	64 (1996)	By annual budget	Vehicle fee, direct; fuel levy through consolidated fund	Yes	Auditor General
New Zealand	$0.065 per liter	580 (1996–97)	By annual budget	Direct deposit	Yes	Auditor General
Malawi	$0.065 gas; $0.074 diesel	16.0 (1998 est.)	By board	Direct deposit	Yes	Independent audit
Romania	25 percent ex refinery price	250 (1997 est.)	No mechanism	Direct deposit	Yes	Independent audit
Russia[c]	25 percent ex refinery price	640 (1993)	By annual budget	Through consolidated fund	Yes	Ministry of Finance
South Africa	None[e]	150 (1995)	n.a.	n.a.	Yes	Auditor General
United States[c]	$0.032 gas; $0.048 diesel	21,000 (1995)	By annual budget	Road fund is line of credit	Yes	Auditor General
Yemen	$0.004 per liter	7.0 (1997 est.)	By annual budget	Direct deposit	Yes, in draft	Auditor General

n.a. Not applicable.

Note: In addition to the road funds shown in this table, there are national road funds and earmarking devices in Belgium, Luxembourg, the Netherlands, Switzerland, Benin, Central African Republic, Chad, Kenya, Madagascar, Mozambique, Rwanda, Senegal, Sierra Leone, Tanzania, Zambia, Zimbabwe, Argentina, Guyana, Honduras, Azerbaijan, Czech Republic, Georgia, Kazakhstan, Lithuania, Mongolia, Slovak Republic, Turkmenistan, Ukraine, and Uzbekistan. New road funds have recently been set up in Jordan and Namibia, and Armenia is currently establishing one. The Swedish government is also considering whether the Swedish National Road Administration should in future be financed solely through vehicle license fees and a fuel levy of about $0.08 per liter.

a. Excluding general budget allocations.

b. Standard government rules and regulations.

c. National or federal road fund.

d. Road fund is not a separate account but a mechanism for financing the regular road budget.

e. Expected to be $0.044 when fuel levy is reintroduced.

levy for the South African road fund was re-instated in early 1998.

Characteristics of Existing Road Funds

The administrative and financial characteristics of existing road funds can be examined by looking at 11 separate areas. The detailed management of the road funds in Japan, New Zealand, and the United States are summarized in annex 4.

Legal Basis

Most road funds are set up under basic legislation or under ministerial and presidential decrees. Decrees are fairly common in West Africa and have also been used in Mozambique and Yemen. Only a few road funds have been set up under the finance act, usually in former anglophone countries (such as Lesotho, Tanzania, and Zambia). In several countries inflexible legislation has caused difficulties. For example, if the legislation sets cost-sharing arrangements and the level of the fuel levy, these specifications cannot be changed without

---

Box 8.2 Urban road funds

HONDURAS: In 1996 the city of San Pedro Sula set up a road fund under a municipal decree. The road fund is managed by a board that consists of a chairperson (the mayor), a vice chairperson, a secretary, and nine voting members from the private sector. The voting members represent the business community, the engineering profession, organized labor, the road transport industry, and the press. They are appointed by the mayor, who chooses from three names nominated by each organization represented on the board. The board hires its own staff. The revenues for the fund come from local improvement taxes, vehicle registration fees, traffic fines, parking charges, and other miscellaneous revenues that are channeled through the municipal budget. Revenues come to about $2 million per year. Maintenance is the first charge on the road fund, but the fund can also finance other road expenditures. There is an independent annual audit.

LATVIA: In 1994 the Cabinet issued a resolution to enable the city municipalities of Riga, Daugavpils, Liepaja, Jelgava, Ventspils, and Jurmala to set up road funds to finance municipal streets. Each fund is managed under an advisory board appointed by the municipality. The head of the Ministry of Transport's regional road service is a member of the board. The revenues for the road funds include transfers from the national road fund—30 percent of the annual vehicle license fee and 27 percent of the fuel levy. Income from the vehicle licenses is distributed among the urban municipalities on the basis of vehicle registrations,

while the fuel levy is distributed on the basis of weighted road lengths (weighted to reflect surface condition and traffic volume). These revenues are supplemented by allocations from the municipal budget and miscellaneous revenues. Funds can be used for maintenance, rehabilitation, improvement, and related expenses. The annual budget and annual report must be published in the local press.

SOUTH AFRICA: Urban roads in the declared Metropolitan Transport Areas are financed through local government rates and grants made from an Urban Transport Fund administered by a subcommittee of the South African Roads Board. The original intention was to partially support the Fund with revenues collected by applying road congestion charges in urban areas. But these charges were never introduced. Instead, the Urban Transport Fund received money from the road fund (in 1986–87, $30 million was transferred), and it is currently financed entirely through a central government grant amounting to about $15 million per year. Money from the Fund is used to finance urban transport plans and infrastructure improvements that assist public transport. The Fund's main responsibilities are being devolved to the nine provinces. In the future, the national Urban Transport Fund will shrink since each province will establish its own fund that it will support with its own revenues. The national Urban Transport Fund will be used to assist in the development of national transport policy, standards, and guidelines, and will finance and administer national demonstration projects.

---

amending the basic legislation (this is a problem in Romania, for example). Likewise, if certain key constituencies are omitted from the board, it is very difficult to add them at a later stage—a problem in Mozambique and Yemen. The Yemen road fund has attempted to solve this problem by inviting two nonvoting members to join the board to represent the Chamber of Commerce and Ministry of Transport.

Oversight Arrangements

Many road funds are overseen by a public-private board made up of representatives from key government ministries and the main road user groups. The rest are overseen primarily by government departments or ministries. The growing number of representative boards generally have members nominated by the organizations they represent; nongovernmental members, including representatives of the business

community, road transport industry, farmers, and the professions; and one member appointed as chairperson or an independent chairperson appointed by the responsible minister (sometimes after consultation with the board). The representative boards attach much importance to accountability. They will often erect signs to inform the public that "these road works are being financed by your money through the roads board" (such as in Yemen and Zambia) or publish their accounts in the press (as in Kenya, Latvia, and Zambia). Board members are normally appointed for three to four years and may be eligible for re-appointment.

Type of Entity

Most road funds are becoming separate agencies, particularly when funds are channeled to several different road agencies and are headed by a secretary or chief executive appointed by the board. But a number are still managed by a division in the main road agency, creating an obvious conflict of interest. Management by a group in the treasury or maintaining the road fund as simply a bank account (as in Ghana before 1997) is now rare.

Staffing

There are nearly always some staff assigned to manage the road fund. They are normally appointed by the chief executive. Transfund employs 35 staff to manage an annual turnover of about $580 million, the Latvian road fund employs three staff to manage an annual turnover of about $64 million, the Russian federal road fund employs 38 staff to manage an annual turnover of about $640 million, and the South African Roads Board employs the equivalent of 10 to 12 full-time staff to manage a turnover of about $150 million. The danger is that the road fund may recruit too many staff (at one stage the road fund in Central African Republic employed more than 50 workers) and that the staff may not have relevant financial qualifications. Thus some road funds have regulations requiring that administrative costs remain under 5 percent of revenues collected through the road tariff (for example, the regulations in Zambia limit the costs of the secretariat to no more than 5 percent of revenues). To ensure they can recruit staff with the necessary financial qualifications, some road funds (as in Ghana) employ workers as consultants, while others (as in Lesotho) have contracted out financial management of the road fund to a firm of accountants.

Qualifying Expenditures

Some road funds finance only maintenance (particularly in Latin America); some finance mainly maintenance but also permit a limited amount of rehabilitation, upgrading, and new works (countries sometimes put a cap on such expenditures); while others finance all road expenditures. There are also some special cases. The road fund in South Africa finances only national roads, that in Argentina only provincial road expenditures on a 50-50 cost-share basis, and the U.S. Federal Highway Trust Fund finances primarily capital works on the federal-aided network (most states have their own highway or transportation trust funds that finance the other roads under their jurisdiction).[2]

When a road fund finances all road expenditures, there is a danger that new works will drive out maintenance. For this reason legislation may state that the road fund is being set up to "provide funds for financing the maintenance, rehabilitation, and improvement of all classified roads," but may then specify that maintenance is the first priority and that "a limited amount of road upgrading, rehabilitation, and minor works" can be financed, "but only after all road maintenance requirements have been met." Malawi has gone one step further, placing a cap of 10 percent on the amount of road fund revenue that can be spent on new works. The Korean road fund restricts the items that can be financed through the road fund. For example, grants made to the provinces do not cover the costs of land acquisition. This is a sound idea, since the costs of land acquisition can be manipulated and the road agency closest to the road works is in the best position to ensure that land costs are not artificially inflated.

Source of Revenues

Most road funds derive their revenues from taxes and charges on fuel, vehicle license fees, international transit fees, and fines for overloading. A number of countries have also introduced procedures to ensure that the fuel levy applies only to transport fuels (Japan, New Zealand, the United States) or that non-road users—generally users of diesel—are compensated for having to pay the fuel levy (Latvia, Mozambique). Some road funds also derive part of their revenues from non-road-related taxes (such as enterprise taxes in Georgia, special excise taxes in Korea, and vehicle sales taxes in Russia and the United States) or from the general budget (South Africa and Honduras are the only countries that derive all their revenues from the general budget). Most of these road funds attempt to ensure that non-road users do not have to pay the diesel levy.

Fuel Levy

The fuel levy is generally specified as a discrete amount per liter or as a percentage of the ex-refinery or wholesale price.[3] This ensures that the fuel levy can be clearly separated from the import duties, sales taxes, and excise taxes that go into the consolidated fund. Japan and Korea are exceptions. In Japan 25 percent of the gasoline tax is paid into the national road fund, while in Korea 67.5 percent of the taxes on gasoline and diesel are paid into the road fund. This old fashioned earmarking can lead to fiscal inflexibility unless the percentage is adjusted each time the fuel tax rates are changed.

Adjusting the Charges

Since most road funds still set the road tariff under the government's tax-making powers, they use the normal budget process to adjust the charges paid into the road fund. The oversight board often determines the level of the charges and recommends them to the cabinet or minister of finance for inclusion in the budget (this system is used in Lesotho, Yemen, and Zambia). In Japan the Ministry of Construction prepares the five-year road improvement program and, after endorsement by the Road Council, the Ministry of Finance sets the appropriate tax rates. In New Zealand the Ministry of Finance sets the cut-off benefit-cost ratio for all new road works, Transfund prepares a national roading program based on this cut-off ratio, and the Ministry of Finance then adjusts the gasoline levy to ensure that the program is fully funded. In Malawi the board sets its own charges, submits them to the minister of works, and, provided the minister is satisfied that the charges are consistent with the government's fiscal targets, the charges become effective and are published in the gazette. Since the tariff is no longer collected under the government's tax-making powers, there is no earmarking. Namibia has introduced similar procedures. In the United States charges are set as part of the overall budget debate.

Surprisingly, some road funds have no mechanism for adjusting charges other than amending the basic road fund legislation (as in Georgia and Romania), or they have a mechanism that involves so many ministries that the charges cannot easily be changed (as in Mozambique).

Depositing the Revenues

Some road funds are simply lines of credit—the treasury credits certain taxes and charges under a single heading in the budget statement (as in Japan, New Zealand, and the United States). The treasury often pays interest on the balance of the account to recognize that it is a separate fund. In some other countries (particularly those with old-style road funds, such as Mozambique and Tanzania) the revenues are first deposited into the consolidated fund and are then transferred to the road fund.

This system often causes problems. A growing number of countries (Ghana, Malawi, New Zealand, Romania, Yemen) have therefore introduced legislation permitting the road-user charges to be deposited directly into the road fund. In others (Lesotho, Sierra Leone) deposits into the consolidated fund is treated as a paper transaction, while cash is deposited directly into the road fund.

Financial Regulations

These spell out how the road fund is to be managed. Some road funds have no special financial regulations or simply rely on general government audit procedures. Others have regulations that are unduly cumbersome and do not always provide sufficient coverage (that is, do not specify who is entitled to receive money from the road fund, how withdrawals are to be authorized, and what the auditing arrangements and annual reporting procedures are).

Auditing Arrangements

Most road funds are audited by the auditor general's office or by private auditors—often appointed or recommended by the auditor general—who may be local or international. Current practice tends to favor international auditors, although this may be due to the influence of the international donor community. The audit normally includes a site inspection of selected schemes financed through the road fund. Auditors' comments on the accounts are frequently included as part of the road fund's annual report.

Problems with Conventional Road Funds

There are many road funds in Africa and Eastern Europe, and a few in Latin America and the Middle East, that were set up during the 1970s and 1980s. Recent reviews have helped to identify what works and why.

Most of the early criticism of road funds focused on the fiscal objection to earmarking and to the operation of extra-budgetary funds (McCleary 1991). These objections have been well-stated elsewhere and can be dealt with by designing the road fund to minimize the adverse fiscal consequences of earmarking and produce efficiency gains that more than offset any remaining fiscal inflexibility. But the most telling argument against road funds—at least those designated as conventional or first-generation road funds—is that they simply do not work.[4] They do not provide a stable flow of funds and they do not strengthen financial discipline.

The above assertion is proven by the audit reports on the early road funds and by the fact that many road funds are not even subjected to a separate technical and financial audit or, when they are, the results are not made public. The available audit reports point to four generic problems: difficulties collecting the revenues attributable to the road fund; the making of unauthorized withdrawals from the road fund—normally referred to as "raids"; payment for goods and services that are either substandard or never delivered to the road agency; and poor financial management of the accounts.

Audit reports regularly mention problems encountered with the revenues designated for the road fund:
• Payments made to district treasuries and commercial banks failed to appear in the road fund account.
• Documentation was insufficient to validate whether all fuel levies had been paid into the road fund account.
• The revenues were collected, but customs disregarded the legislation and paid the proceeds into the consolidated fund.
• Funds were collected by the oil company, but significant amounts disappeared before being deposited into the road fund—funds disappeared between collection by the oil company and deposit into the ministry of works account, and again between the ministry of works account and deposit into the road fund account.

Audit reports regularly mention cases involving unauthorized withdrawals from the road fund:
• Money was taken from the road fund to pay civil service salaries.
• Although the funds were intended for road maintenance, they were instead used to purchase vehicles and refurbish the state house and parliament building.
• Funds were used to pay for items that did not qualify, including hotel bills, construction of houses, refurbishment of offices, utility bills, and gratuities.
• Although the funds were intended for maintenance, they were instead used to finance capital works.

The audit reports also mention numerous instances in which payment was made in full for substandard work and even for goods and services that were never delivered:

• Payment was made for vehicles and materials that were never supplied to the road agency (they may have, however, been delivered to others).
• An inspection of some rehabilitated roads showed they were still in poor condition.
• Funds were used to lay 40 mm of premix asphalt concrete on a newly constructed carriageway. But an audit site visit could not trace the newly constructed carriageway, and the street was found to be in "pathetic" condition.
• A contractor was paid to construct three culverts—they could not be found during a site visit.

Finally, the audit reports are full of instances pointing to poor record keeping and weak financial management, largely attributable to lack of guidelines or financial regulations for control and management of funds. Likewise, some road funds do not have any accounting staff and do not even keep a ledger account to record deposits and withdrawals. The following issues are regularly mentioned in audit reports:
• Inability to certify accounts because of inadequate record keeping.
• The absence of bank reconciliation statements to support the cash shown in the balance sheet and held in the bank account.
• The absence of reliable records to show what work has been done.
• Money shown as having been paid to municipalities, but without records to verify that the funds were received.
• Instances of overpayments, lapses in procurement procedures, and payment without supporting vouchers.

Road funds also experience other problems that tend to be associated with poor design or difficult country conditions, rather than with weak governance and poor financial management. These include:
• *Legal problems.* Road funds often lack a clear legal basis—usually because the legislation is prepared too quickly or without sufficient care and attention. The legislation may then carry inconsistencies, ambiguities, or be too rigid.
• *No mechanism for objectively allocating funds among different road agencies.* Many road funds have no objective procedures for meting out funds to the different road agencies. As a result, the allocations tend to be erratic and subject to political whim.

• *Insufficient road fund coverage.* When the road fund finances part of the cost of road works, leaving the balance to be paid from the government budget, it often becomes even more difficult to get hold of the agreed budget funds. Other ministries argue that the road sector has already received its funds and that their sectors now deserve funding.

• *Oil companies withhold payment.* This usually happens when the government is in arrears with its fuel payments. The oil companies withhold payment of government sales and excise taxes and may also stop paying the fuel levy into the road fund.

• *Excessive road fund revenues.* If the road tariff is set too high, excessive funds may accrue. Other ministries may then raid the fund or bring pressure to have it closed. This happened in the Philippines (its road fund was set up in the early 1950s, modeled on the U.S. Federal Highway Trust Fund) in the early 1970s and in South Africa in 1988. About one-third of the pump price of fuel was earmarked for the South African Road Fund. When the levy was abolished in 1988, more than $0.15 per liter was earmarked. This high levy caused the Road Fund to build up a large surplus, raising concern that the South Africa Roads Board might start building uneconomic roads. The levy was eventually abolished.

Setting Up a Commercially Managed Road Fund

The above reviews identify the issues that must be borne in mind when designing a new road fund or restructuring an existing one. If these issues are addressed up front, the road fund is likely to be supported by the ministry of finance and the IMF (boxes 8.3 and 8.4). Three groups of questions must be asked when designing such a road fund:

Strategic questions:
• What kind of road fund is needed, that is, which parts of the road network will it finance?
• What kind of legal basis should it have?
• What sort of oversight arrangements (or governance) does it need?
• How should the road fund be managed (that is, through what kind of agency)?
• Which expenditures should the road fund finance?

Technical and policy questions:
• Where should the revenues come from and how should they be deposited into the road fund?
• How should the road tariff be adjusted?
• How can non-road users be exempted from paying the diesel levy?
• How should funds be allocated among the different road agencies entitled to receive money from the road fund?
• What sort of cost-sharing arrangements should be used on local government roads financed through the road fund?
• How should funds be disbursed to each road agency?

Operational questions:
• How should day-to-day management of the road fund be organized?
• What sort of financial rules and regulations are needed?
• How should the road fund be audited?

Strategic Questions

What kind of road fund is needed? This question is the most important of all. There are four broad options and numerous variations on these options. The first is to set up the road fund to finance only national roads (as in South Africa and the United States). This is not a particularly good arrangement unless road-user charges can be confined to the national road network or each local jurisdiction has its own road fund (as in the United States). Most revenues normally come from a fuel levy that is paid by all road users. The revenues should therefore be used to finance all roads. It makes little sense to have only part of the road network well-financed. Furthermore, when setting up a new road fund, it is important to have as much support as possible—which usually means providing some funds for urban and rural roads to win the support of local governments.

The second option is to set up the road fund to finance all qualifying expenditures on the national road network and to support maintenance and improvement of local government roads through grants (as in Korea, Latvia, and Russia). This option has several attractions. The road fund supports all roads, but can confine transfers to local governments to what it can afford after the demands of the national road network have been met.

Box 8.3 Earmarking versus commercialization: A World Bank view

Commercialization is not the same as earmarking general budget revenues as a means of capturing more of the government's overall budget for the road sector. World Bank staff reject that approach, which was unfortunately a characteristic of many road funds in Latin America and Africa in the 1970s and early 1980s. Straightforward earmarking has never worked. This is clearly spelled out in several World Bank reports. Furthermore, commercialization is not necessarily adopted in countries with dysfunctional budgetary systems. On the contrary, it is often adopted as part of the process of redefining the role of government, as in New Zealand, which is normally held up as an example of a country with sound fiscal management.

Commercialization calls for bringing roads into the marketplace, putting them on a fee-for-service basis, and managing them like a business. Four main principles underlie this concept:

• Road users pay for usage of roads through an explicit road tariff that must be clearly separated from the government's general taxes. It usually takes the form of a two- or three-part tariff: an annual vehicle license fee that charges for access to the road network (sometimes supplemented by a heavy-vehicle license fee), a road maintenance levy added to the price of fuel that charges for use of the road network, and, where feasible, a congestion charge to manage congestion.

• Introducing the above road tariff must not abstract revenues from the consolidated budget. The ministry of finance is generally invited to convert the existing allocations for road maintenance into an equivalent fuel levy, but that is all. Any additional revenues must come from extra payments by road users. That is part of the objective—road users pay for using the road network, they know that they are paying, and they are thus encouraged to demand value for money.

• The proceeds from the road tariff are deposited into a road fund managed by a board that includes representatives of road users and the business community. At least half of the board members generally come from outside the government and are nominated by the organizations they represent. The chairperson is independent. This structure creates a form of surrogate market discipline. Board members represent the people who are paying for the roads and they thus have a strong vested interest in seeing that they are not overcharged and that the money is well spent.

• Finally, the board must have a small secretariat to manage the funds, published legal regulations should govern the way the funds are managed, and the auditor general's office or private sector auditors appointed by the auditor general must carryout independent technical and financial auditing. This system is referred to as commercial management.

Additional elements are that the fund should ideally support maintenance of all roads (including cost-sharing with local governments and communities), responsibility for different parts of the network should be clearly assigned to a competent road authority, and the road authorities should introduce sound business practices. Indeed, once you have a representative road board, the members usually insist that road agencies operate along commercial lines.

Some road funds already have procedures in which the road tariff no longer forms part of the consolidated fund. Instead, the road sector has been set up as a road public utility. The board sets its own charges, submits them to the minister responsible for transport, and, provided they are consistent with the government's overall fiscal targets, publishes them in the government gazette. The road tariff is thus no longer collected under the government's tax-making powers, and the revenues are no longer earmarked taxes. Instead, the revenues are collected under contract by the oil companies and government departments, and they are deposited directly into the road fund. Some of the strongest supporters of the above system are the ministries of finance. They see it as making road financing more transparent and as tightening financial discipline in the road sector. Ministries of works are less enthusiastic, since it imposes on them a large measure of (unwelcome) financial discipline.

Source: Extracts from a memo from the World Bank Roads Adviser to the IMF Senior Managing Director.

The third option is a variation on this system and is exemplified by the Latvian road fund. The legislation provides for grants to regional municipalities but also permits each municipality to set up its own road fund to manage the grant. The municipality must establish a management board, include the regional representative of the transport ministry on the board, and publish its annual accounts in the press—quite a novel approach.

The final option is to set up the road fund to finance all roads, but to finance local roads on a cost-share basis (as in Ghana, New Zealand, and Zambia). In many ways this arrangement is ideal, although it complicates management. If funds are going to be chan-

Box 8.4 Commercialization of roads: The view of Jordan's Minister of Finance

"To start with, we need to think of road funds which reflect a desire to pursue an Agency Model of service delivery for roads under which the management of roads is commercialized with expenditure on roads being financed from user charges which should reflect the view that 'roads should be managed like a business, not like a bureaucracy'. My Ministry has no objection to the above 'user-pay' or 'fee-for-service' principle, but this involves not only the establishment of a road fund, but a rethinking of our whole approach to the road sector which often leads us to conclude that we need to establish a road fund. Let me mention in some detail the main concerns of my Ministry regarding the establishment of a road fund.

First, we want assurance that the establishment of the road fund is part of a longer term strategy to commercialize the road sector, and that it is not simply a means of avoiding strict budget discipline.

Second, we expect the road fund to be dedicated to maintenance. We must make sure that we maintain what we have, before starting to build anything new.

Third, we expect to see the road fund as a purchaser, not a provider of services. It should be a separate agency with a clear mission statement, transparent objectives, physical output indicators and it should ideally work within an envelope of total input costs.

Fourth, we expect the road fund revenues to come only from road-user charges, not from any earmarked taxes. That would not prevent the government from topping up the road fund from the consolidated budget, but topping up would only be done on a discretionary basis.

Fifth, the most fundamental requirement of all. The user charges going into the road fund must not take revenues away from other sectors. We would like to see a clean break between the tax revenues which belong to the consolidated budget, and the user charges which belong to the road fund. The only existing revenues which should go into the road fund, must be confined to what is already allocated for roads through the annual budgeting process.

Sixth, we expect to see the road fund managed by a strong and independent management board which should include private sector interests—both road users and the business community—and should be genuinely free from any vested interest groups.

Seventh, we expect the management of the road fund to be handled by a secretariat and to employ commercial accounting systems and to have annual performance targets.

Eighth, we want to see a fair degree of cost recovery through the user charges. We look in the long-term for a road public utility which does not receive any government subsidy.

Ninth, we cannot escape from the fact that fuel is a convenient tax handle from the point of view of fiscal policy. That will inevitably put a burden on the road fund administration to explain to the public why all fuel price increases are not equal.

In brief, we are perfectly willing not only to consider establishing a road fund, but even to actively help to get it established. However, you have to assure us that what we are establishing is the right kind of road fund and that it will be based on sound fiscal principles."

Source: Text of speech given by His Excellency Suleiman Hafez, Minister of Finance, Jordan, June 3, 1997.

neled to all roads, there must be an approved national roads program, agreed cost-sharing arrangements with local governments, agreed procedures under which local governments will manage their share of the road fund, and appropriate financial and technical auditing procedures. But once these procedures are in place, it is probably the best type of road fund to have.

What kind of legal basis should it have? A road fund can be established in one of three ways: through the finance act, a device often used in countries with British-style legislation; by issuing a ministerial or presidential decree; or by passing special-purpose legislation. The first two options are the simplest. The finance act often gives the minister of finance power to open a special

account for a designated purpose, while many countries have legislation that permits the minister, president, or cabinet to publish a decree setting up a road fund. This legislation can be as simple as publishing a legal notice in the government gazette or may involve passing a parliamentary resolution or a cabinet decree. Ideally, the legal notice should state that a special road fund account is being opened, why it is being opened, what will be the source of revenues, and how the account will be managed. But in many cases it merely needs to state that a special road fund account is being opened and that the detailed arrangements for managing the road fund will be published in a separate notice in the gazette (box 8.5).

There are two drawbacks to the above procedures. First, they provide only a temporary basis for the road

Box 8.5 **Notice used in Lesotho to open a special account (road fund)**

LEGAL NOTICE NO. 179 OF 1995
Finance (Roads Fund) Notice, 1995

In exercise of the powers conferred on me by Section 16A of the Finance Order, 1988, I

DR. MOKETSI SENAOANA

Minister of Finance and Economic Planning make the following notice:

Citation and Commencement

1. This Notice may be cited as the Finance (Roads Fund) Notice 1995 and shall come into operation on the date of its publication in the gazette.

Establishment of Special Fund

2. There is established a Special Fund to be known as the ROADS RELIEF FUND.

Dr. M. Senaoana
Minister of Finance and Economic Planning

fund and should be accompanied by a sunset clause (that is, once it has been decided to set up the road fund on a permanent basis, basic legislation should be passed). And second, since the road tariff is collected under the government's tax-making powers, the revenues have to be paid into the consolidated fund and then transferred to the road fund, which introduces delays and increases the risks of diversion. Some decrees, however, provide for directly depositing revenues—as in the road fund of Yemen, the former Ghana road fund, and some of the West African road funds—and payment into the consolidated fund does not necessarily mean that the cash has to be deposited into the treasury account. The advantages of using existing legislation are that the road fund can be set up quickly and there is time to iron out early problems before passing basic legislation.

The third option is to establish the road fund under new legislation. This provides a much firmer basis for the road fund, although it takes longer and requires a committed minister willing to take the bill through parliament. When contemplating new legislation, it is important to decide on what needs to go into the legislation and what can be put into supplementary legal regulations (to shorten the act and make it easier to

revise the operating modalities from time to time) and which mechanism will be used to revise the road tariff.

The choice of mechanism must be built into the legislation and made consistent with the country's constitution and the finance act. There is little point to passing new legislation if the road tariff will continue to be treated as if it were part of the government's tax-raising mechanism. Annex 5 is an example of the full text that might be used to set up a road fund under new legislation, and annex 6 provides an example of the regulations that might be used to set up a road fund under existing legislation.

What sort of oversight arrangements does it need? Oversight arrangements are one of the most important design elements. Oversight is normally provided by appointing a board either to advise the minister on management of the road fund or to manage the road fund directly through an executive board. Although it is better to have an executive board, advisory boards have their place and can be highly effective provided they have the right membership. The board should be made up of about 9 to 12 members. Boards that are too large tend to become unmanageable, while those that are too small have difficulty capturing all the key constituencies.

The road fund can also be managed by a subcommittee of a larger road management board (as in Malawi and South Africa). The board should be able to appoint subcommittees to help with its work, and the subcommittees should be able to include people who are not members of the main board. Even the main board should be able to invite outsiders to advise on special topics.

The board should comprise representatives of the key constituencies with vested interests in well-managed roads (see box 8.6). For the central government this usually means representatives from the ministries of works, transport, finance, and agriculture, and representatives from local government. The nongovernment members usually include representatives from the chamber of commerce, the road transport industry, farmers (both commercial and small-holder), and the professions (like the institute of engineers). Local governments should also be represented on the board if the road fund intends to channel funds to them. There may need to be separate members representing large and small local governments.

Ideally, half or more of the members should be non-government or local government representatives. The director of roads should not be a member of the board, but should attend board meetings. Members should be nominated by the constituencies they represent, rather than being selected by government officials or the responsible minister, and should be formally appointed by the minister, president, or cabinet. The names of board members should be published in the government gazette. Appointment should be for a period not exceed-

---

Box 8.6 Membership of oversight boards in Latvia, Malawi, New Zealand, and South Africa

LATVIA: STATE ROAD FUND. The State Road Fund was established under a cabinet decree in 1994. Management of the Fund is overseen by the State Road Fund Advisory Board. The Board has 13 members: six represent the central government (Minister of Transport, Director of Roads, Manager of State Road Fund Division, Director General of Road Administration, Director of Road Safety Administration, and Ministry of Environment and Regional Development); two represent local governments (Riga City and union of local governments and municipalities); and six represent civil society (public transport association, automobile society, road builders, automobile dealers association, and Riga Technical University). The six members representing the central government are de facto ex officio members; the remainder are nominated by the organizations they represent. The Minister of Transport acts as chairperson, and the manager of the State Road Fund Division acts as secretary of the board. The Fund is managed by the State Road Fund Division within the Road Administration, which has a staff of three.

MALAWI: NATIONAL ROAD FUND. The National Road Fund was established under the National Roads Authority Act in 1997 and is managed by a subcommittee of the main National Roads Authority Board. The main Board has 13 members: seven drawn from organizations representing road users, farming interests, the business community, local government, and the National Road Safety Council; three representing the public interest; and three ex officio members representing the Ministries of Works, Local Government, and Transport. Members are appointed by the Minister of Works. The seven members representing organizations of road users are nominated by those organizations. The ex officio members are the respective permanent secretaries or their designated representatives, while the three people representing the public interest are nominated by the Permanent Secretary of Works. The chairperson is appointed by the Minister of Transport from among the members of the board. The deputy chairperson is elected by the board from among its members. The board can establish subcommittees and co-opt nonvoting advisers to assist with the work of the Authority. The Road Fund is managed by one of these subcommit-

tees. The board designates one of the employees of the National Roads Authority to act as secretary to the Board. The National Roads Authority employs a small staff of three to five people to manage the Road Fund.

NEW ZEALAND: NATIONAL ROADS FUND. The National Roads Fund was established initially under the National Roads Act, 1953 and amended in 1989 and then in 1995. It is managed by the board of Transfund, which was set up under the Transit New Zealand Amendment Act, 1995. The board consists of five people: two representing Transit New Zealand (either employees or members of the Roads Authority), one representing local government, one representing road users, and one representing an aspect of the public interest not represented by the other members of the board. The members are appointed by the governor-general on the recommendation of the responsible minister following consultation with people from the land transport industry and elsewhere. The chairperson and deputy chairperson are appointed by the minister from the existing board members. The current chairperson is the representative for local government and the deputy chairperson is from academia. The remaining three members are the former chairperson of the New Zealand Road Transport Association, chairperson of Transit New Zealand Authority, and a further member of the Transit New Zealand Authority who is also a director of the Port of Marlborough. The board has a full-time secretary. Transfund currently has 35 staff, including a CEO appointed by the board. The CEO appoints all other staff.

SOUTH AFRICA: NATIONAL ROAD FUND. The Road Fund was set up under the National Roads Act, 1935, and is currently managed by a subcommittee of the South African Roads Board. The subcommittee consists of one member of the main Board who acts as chairperson, the chief director (roads), the director of financial planning and administration, the departmental accountant, and a co-opted accountant from the private sector. The main Board member always acts as chairperson of the subcommittee. The road fund is managed by staff in the office of the chief director (roads). Management of the road fund currently absorbs the equivalent of about 10–12 full-time staff per year.

ing three to four years, and members should generally be eligible for re-appointment for at least one more term.

There should be an independent chairperson. The procedure in Zambia, where the board elects its own chairperson, is unusual, although it has worked well. Jordan is now proposing that board members elect the chairperson. Normally, the minister either appoints one of the existing board members as chairperson or appoints an outsider after consultation with the board. The vice chairperson is often elected by the board. The board must have a clear (published) terms of reference spelling out its role concerning public support for more road spending (that is, its outreach program), which expenditures the road fund can finance, how it is expected to manage the road fund, its relationship with the minister, and the basis on which the minister can issue directives (box 8.7).

How should the road fund be managed? There are several ways of managing the road fund. If one road agency is responsible for managing the entire road network, the road fund and the road network can be managed by the same board without creating any conflict of interest (as in Sierra Leone). Likewise, if the road fund finances only the main road network, it can be managed by the board that manages the main road network without creating any conflict of interest (as in South Africa). Otherwise, there is a danger that the road fund will attend to the needs of its own roads first, and channel only what is left over to the other road agencies. In such cases it is better to establish a separate road fund board to channel funds in an even-handed way to the appropriate road agencies. This system is now at work in Ghana, Lesotho, New Zealand, Malawi, and Zambia.

Not many staff are needed to manage the road fund. They normally collate the road programs prepared by the various road agencies, review and consolidate them into the approved national road program, define the financial procedures to be followed by the agencies receiving money from the road fund, allocate funds to support the approved programs, disburse funds to the road agencies, and then audit results ex post. The staff should also audit the systems and procedures used to prepare the road program and control expenditures— as in New Zealand and the United States—and should

manage the day-to-day affairs of the road fund. Based on the figures for the Latvian, New Zealand, Russian, and South African road funds, it should not take more than about five staff to manage an annual turnover of about $100 million, and 30 to 35 staff to manage an annual turnover of about $500 million.

Which expenditures should the road fund finance? Most road funds have been set up primarily to finance routine and periodic maintenance of the existing road network. To ensure that spending on new works does not drive out maintenance, the road fund can be set up as a road conservation fund, which can finance only road maintenance (as in Latin America), or the matter can be dealt with in the legislation (or associated legal regulations). In the latter case the legislation (or regulations) should state that maintenance has first claim on road fund revenues and that, only after all maintenance has been fully funded, may the balance be spent on rehabilitation and new works. Spending on road safety and administration of the road fund should also have first claim on revenues, though spending on administration should be subject to a cap to prevent the road fund from simply becoming an employment agency.

Most road funds also finance road rehabilitation or provide counterpart funding for rehabilitation programs financed by donors. A few finance new investment. It is important that these priorities be clearly stated and prioritized in the legislation. Alternatively, they can be spelled out in the legal regulations so that they can be revised from time to time to reflect changing circumstances. If the road fund does finance new investment, it may be important to put a cap on the amount. If it does not finance new investment, the legislation (or regulations) should state explicitly that new investment will continue to be financed through the government's development budget.

Technical and Policy Questions

Where should revenues come from and what deposit procedures should be practiced? The only source of revenues should be the charges making up the two- or three-part tariff, namely, vehicle license fees, supplementary heavy-vehicle fees, international transit fees, the fuel levy, fines for overloading, and any charges imposed to internalize the costs of road congestion

---

Box 8.7 Terms of reference for road fund boards in Latvia, Malawi, New Zealand, and South Africa

LATVIA: STATE ROAD FUND. The Fund is overseen by the State Road Fund Advisory Board, established by statute. The detailed functions of the board are to:
• Review general strategies for revenues and expenditures, together with the proposed annual expenditure program.
• Periodically review collection of revenues for the Road Fund.
• Review the planning and use of funds allocated to municipalities.
• Inform the public about the work of the Road Fund.
• Report all matters reviewed by the board to the responsible minister.

MALAWI: NATIONAL ROAD FUND. The main functions of the road fund subcommittee are to raise funds to ensure that public roads are fully maintained and rehabilitated. The detailed functions of the subcommittee are to:
• Review the annual road programs prepared by the road agencies and consolidate them into a national program, submitted to the minister for approval.
• Determine the allocation of financial resources required by road agencies for maintenance, rehabilitation, and development of public roads.
• Recommend to the minister appropriate road-user charges, fines, penalties, levies, or any other sums to be collected under the act and paid into the road fund.
• Disburse funds or authorize payment of funds to contractors only after it has been certified in writing that the work has been carried out to the required standard.
• Prepare, publish, and submit to the minister audited annual accounts for the fund and also submit, at such intervals as the minister shall provide in writing, reports

and financial statements regarding operations of the Authority, the board, and the Fund.

NEW ZEALAND: NATIONAL ROADS FUND. The principal task of the board is to allocate resources to maintain a safe and efficient roading system. The detailed functions of the board include:
• Approve and purchase a national roading program that prioritizes funding consistently on the basis of expected national benefits for a given cost.
• Pursue efficiency in delivering roading and alternatives to roading through contestability and through promoting enhanced administrative and technical systems and processes.
• Establish contracts with road-controlling authorities and regional councils for the delivery of their respective programs.
• Audit all road-controlling authorities and regional councils on a timely basis to provide assurance as to the efficient and effective use of resources.
• Establish the process for evaluating and funding efficient alternatives to the provision or maintenance of roading.

SOUTH AFRICA: NATIONAL ROAD FUND. The Finance Committee advises the board on:
• How to spend the available funds in the most cost-effective manner.
• The annual budget, the five-year road expenditure program, and the overall financing strategies of the board.
• The financing of toll roads, including the timing, amount, and rates on new bond issues.
Since the fuel levy has been reinstated in 1998, the road fund is managed by a new audit committee that advises the board on all financing issues and undertakes performance audits of all activities financed by the road fund.

(parking charges, cordon charges). Revenues should not include any earmarked general taxes (import duties, sales and excise taxes). Miscellaneous sources of revenue, like bridge and ferry tolls, donor funding, and contributions from the consolidated fund, may also be added.

The fuel levy should be specified as a discrete amount (that is, so many cents per liter), or as a percentage of the ex-refinery or wholesale price of fuel (or the equivalent). It should not be specified as a proportion of fuel taxes or other general revenue taxes, since this would turn it into an earmarked tax. When first introduced, the charges must be set to ensure that they do not abstract revenues from other sectors of the econ-

omy. In other words, if the consolidated fund can afford to finance only 20 percent of maintenance requirements, then only that 20 percent may be converted into the initial license fees and fuel levy transferred to the road fund. All additional revenues must come from extra payments made by road users.

The next question is, which mechanism should be used to deposit revenues into the road fund? When the road fund is set up under the finance act, revenues must be paid into the consolidated fund and then transferred to the road fund. But if the ministry of finance agrees, the funds can be deposited into the consolidated fund as a paper transaction, and the actual cash deposited directly into the road fund. Likewise, when the road fund is set up under a decree, the revenues may have to be deposited into the consolidated fund, though some decrees provide for direct deposit into the road fund. This method should be used whenever possible.

When the road fund is set up under new legislation, the act should be drafted to permit the proceeds from the road tariff to be deposited directly into the road fund. The act should also permit the tariff to be collected under contract (even when collected by a government department) and, when feasible, to be collected under competitively awarded contracts (see chapter 7 for examples of contractual arrangements).

How should the road tariff be adjusted? A formal mechanism should be in place for adjusting the road tariff—both upward and downward—to ensure that the road fund generates sufficient revenues to meet approved expenditure requirements but does not generate excessive revenues. The oversight board should have the power to set the road tariff (in the same way that the railways set their tariffs), or to at least recommend tariff levels to the ministry of finance for inclusion in the annual budget statement. Malawi and Namibia are the only countries that have set up public utility–style road financing mechanisms, although New Zealand expects to be operating as a road public utility within about two years.

Under the public utility arrangement the road fund board must set the level of the road tariff. In doing so, it should bear in mind the revenues required to finance the approved road expenditure program and road users' willingness to pay. The board then submits its recommendation to the minister of works or transport and, provided that the minister is satisfied that their proposals are reasonable and consistent with the government's overall fiscal targets, the revised tariff becomes effective and is published in the government gazette.

But most countries still set their charges under the government's tax-making powers. Then, the board (or the responsible ministry when there is no board) recommends the revised charges to the minister of finance or the cabinet, and, once approved, these charges are included in the annual budget statement. The ministry of finance or the cabinet is more likely to accept the recommended changes automatically if they are submitted by a representative board.

When the road fund is first set up, the tariff will usually have to be raised gradually over a period of three to five years. This slow build-up enables the board to show results to its constituents before asking for further increases in the road tariff. Many boards operate extensive outreach programs to demonstrate to their constituents that they are getting value-for-money from the road tariff. Several boards also publish their accounts in the press and plant road signs stating that the road works are being financed by the national roads board. All road fund boards should consider operating similar outreach programs. Finally, while the road fund is building up its revenue base, the balance of required revenues should come from donors or the general budget.

How can non-road users be exempted from paying the fuel levy? When significant amounts of gasoline or diesel are used for non-transport purposes and the fuel levy is perceived to be high, efforts must be made to ensure that nontransport users do not have to pay the fuel levy (these problems apply primarily to diesel). Failure to address this problem may generate strong public pressure to suspend the fuel levy. If there are a few large users (mining companies, power stations), it may be possible to exempt them without encouraging too much avoidance and evasion, although experience is not encouraging. For small users (such as farmers), the only realistic options are: coloring non-transport diesel, testing to ensure it is not being used on the road, and applying stiff penalties for infringements (although used in Finland, the United Kingdom, and the United States, this is difficult to administer); allowing nontransport users to apply for rebates based on invoiced consumption for non-transport uses (Namibia is proposing to use this method, although it requires extensive auditing and is also difficult to administer); or compensating non-transport users for having to pay the fuel levy. The last method is probably the simplest to administer and is used successfully in both Latvia and Mozambique to compensate farmers. Latvia uses the same system to compensate the railways and fishing industry.

How should funds be allocated among different road agencies? A simple and consistent procedure is needed for allocating funds among the different agencies entitled to receive money from the road fund. The procedures must be transparent, fair, related to need, and, where feasible, related to the road agency's ability to generate funds from other sources. There are two basic approaches. The road fund can either allocate the funds using formulas or base the allocations on a direct assessment of need.

A formula-based system usually starts by allocating the funds among the main, urban, and rural road agencies and then goes on to subdivide each allocation among the individual road agencies within each group. The road fund will therefore allocate a certain percentage of its revenues to urban roads and a certain percentage to rural roads, with the remainder going to the main road network. For example, Latvia allocates 27 percent of the annual vehicle tax and 30 percent of the fuel levy to municipalities (both urban and rural), Mozambique allocates 20 percent to urban municipalities, Romania allocates 35 percent to county and communal (village) roads, and Zambia allocates 25 percent for rural roads and 15 percent for urban roads.

Initially allocating the funds among the different road agencies ensures that each gets a fair share of the revenues available, provided the proportions are regularly amended in light of experience. This is an important consideration when strong urban councils bid against weak rural councils for the same share of the road fund. For example, in Tanzania 20 percent of the road fund is set aside to finance road works managed by 17 urban and 84 rural district councils. But the urban councils are generally able to prepare better road programs and have more political influence. Thus three-quarters of the funds have gone to them, mainly to the capital city, Dar-es-Salaam. Another reason for initially dividing funds at the source is that different types of road agencies may use different criteria to establish priorities, reflecting their differing technical capacities. Finally, initially allocating funds gives each group of road agencies an indicative guideline upon which to base their spending plans.

The next step is to separate each allocation among the road agencies in each group. There are two main ways of doing this. Either each road agency must compete for the available resources or the resources are allocated on the basis of network and traffic characteristics. Under the first system the road agencies bid for the funds, a panel evaluates the bids, and then decides which road agency should get what. The bids cover both maintenance and investment programs.

Hungary and Zambia use this system. It is not a particularly good way of meting out funds (some road agencies end up being fully funded, while others get little or nothing), although it does encourage road agencies to put a lot of effort into planning and justifying their road programs.

Under the second system revenues for investment are usually allocated using benefit-cost analysis, as under the first system. The road fund usually issues guidelines on how the investment programs are to be prepared, offers advice on how to compute the benefit-cost ratios, may specify the minimum acceptable benefit-cost ratio, and audits the calculations to ensure they have been carried out correctly. Transfund New Zealand currently has a cut-off ratio of 4.0. Staff from Transfund audit a sample of all benefit-cost calculations, including those for all schemes over $700,000.

Revenues for maintenance, on the other hand, are allocated on the basis of network and traffic characteristics. The road fund generally does so using formulas based on parameters like length of the road network, volume of traffic, and ability to pay. The formulas generally include road length (or lane-km), which may be weighted to reflect estimated maintenance costs on different types of roads (as in Latvia). They may also include vehicle-km or the vehicle population and will often include resident population. Some countries include a term to reflect ability to pay (like Korea, which includes an adverse financial ability index). The U.S. Federal Highway Trust Fund includes a predetermined minimum maintenance allocation (box 8.8).

The direct needs-based approach allocates maintenance funds according to a more careful assessment of network needs (funds for investment are again evaluated using benefit-cost analysis). The methods can be more or less complicated, depending on the technical capacity of the road agencies involved. The simplest way to estimate needs is by using standard unit rates for each routine and periodic maintenance activity according to type of road surface. Each rate is multiplied by each road agency's length of maintainable road in each road class to arrive at the total required maintenance budget. Adjustments may then be made for climatic variations and other factors. South Africa uses this method to estimate multiyear allocations for rural roads in the nine provinces (box 8.9).

Box 8.8 Allocating revenues on the basis of network and traffic characteristics

UNITED STATES: Maintenance funds from the Highway Trust Fund are allocated among states according to the following formula: 0.55*(interstate lane miles/total interstate miles) + 0.45*(vehicle miles on interstate roads/total interstate vehicle miles). The average allocation per state is about 2 percent of total maintenance funds, subject to each state receiving a minimum allocation of 0.5 percent. These allocations cover 90 percent of costs. The General Accounting Office has recently suggested two alternative allocation formulas: distribution based equally on total lane miles and total vehicle miles traveled, and distribution based equally on total lane miles, interstate vehicle miles traveled, and state population.

KOREA: Funds allocated to provinces from the national road fund are divided up on the basis of a formula that takes into account the road length, demand for road space, and provincial ability to pay. The formula is similar to that used in the United States: 0.5*(length of provincial road/total provincial roads) + 0.15*(provincial population/total provincial population) + 0.15*(provincial vehicle population/total provincial vehicle population) + 0.2*(adverse financial ability index for province/total adverse financial ability index for all provinces). The adverse financial ability index = 1/[(local provincial tax revenues + government grants)/local provincial expenditures)].

LATVIA: Funds allocated to municipalities from the national road fund are distributed among 7 city municipalities and 26 regional municipalities (27 percent of the vehicle tax and 30 percent of the fuel levy) on the following basis. Half of the vehicle tax is allocated to the city municipalities and is distributed among them on the basis of vehicle registrations. The balance of the vehicle tax is allocated to the regional municipalities on the basis of weighted (maintainable) road length. The weights crudely reflect maintenance costs relative to traffic levels. The weights currently used are: urban roads of district cities and rural communes = 5; asphalt paved roads = 2; other roads = 1. The fuel levy is distributed to all municipalities on the basis of weighted road length.

A better way to assess maintenance needs is by basing requirements on the output of a standardized road management system. Staff from the road fund will usually have to advise the road agencies on how to operate the systems and then carry out regular audits to ensure that each authority is applying the procedures correctly. The audit should also ensure that the funds are actually spent on maintenance and that the maintenance is carried out to agreed standards. This method requires that road agencies be technically competent, which may not always be the case, particularly at the local government level (box 8.10).

What sort of cost-sharing arrangements should be used? The road fund will have to establish cost-sharing arrangements for local government roads, since it generally finances only part of their maintenance and investment costs. Since local roads benefit not only road users but also adjoining landowners, costs are usually shared between the two (through local property taxes). Cost-sharing arrangements normally consider two factors: the

---

Box 8.9 A simple method for estimating maintenance needs: South Africa

In 1991 the Department of State Expenditure introduced a system for multiyear planning of public expenditures. The arrangements were implemented through a function committee, chaired by the Department of Transport, which developed procedures for equitably allocating funds for rural roads.

To estimate maintenance needs, maintenance is divided into routine maintenance (patching and sealing cracks, maintaining gravel shoulders, maintaining drainage, attending to the road reserve, and maintaining road signs and markings) and periodic maintenance (maintaining bridges, resealing, and making minor road safety improvements). A matrix of unit maintenance rates—for each type of road, traffic condition, and activity group—is then applied to roads under the jurisdiction of each road authority, thus arriving at the total target maintenance requirements.

Roads are classified by type and traffic volume. In the case of local access roads, for which no traffic figures are available and very low maintenance standards are applied, a flat figure of $70 per km is used. After these figures are adjusted to account for environmental conditions, they are then used as part of a structured negotiation process in which the members of the function committee have to agree on the allocation models to be used for each expenditure category and the total allocation for each expenditure category.

Box 8.10 Using a maintenance management system to assess needs: New Zealand

Maintenance requirements for both Transit New Zealand and the local authorities are based on a combination of professional judgment and the outcome of a Road Assessment Maintenance Management system. This system is a computerized pavement management system that includes road inventory data (road condition) and treatment selection for determining work programs based on engineering and economic criteria.

Transfund requires all road agencies wishing to receive financing from the road fund to base their estimated funding requirements on the Road Assessment Maintenance Management system. Staff from Transfund advise the transport authorities on how to operate the system. Road authority requests are vetted on an ongoing basis by Transfund staff, and the Review and Audit Division carries out audits every three years to ensure that each road authority is maintaining minimum maintenance standards and service levels.

---

revenues needed to fully finance the local government's road program and what the local government can afford to finance from its own resources. This affordability is largely related to the extent of fiscal decentralization and depends on which sources of taxation have been delegated to local governments. The revenues required to maintain local government roads are estimated using benefit-cost analysis and formula-based or needs-based methods. Affordability is generally measured in terms of the local property tax base. The larger is the tax base, and hence the wealthier is the local government jurisdiction, the smaller is the amount financed by the road fund. When the tax base is not known, affordability is measured indirectly in terms of population density or number of people served by the road (box 8.11).

When the road fund is first set up it will often finance all road expenditures—or at least all maintenance expenditures—and will move toward an agreed cost-sharing formula over a period of two to five years. Thereafter, the formula should be reviewed and, if needed, amended annually. Typically, the road fund will finance 50 percent of the costs of maintaining urban roads and 60 percent of the costs of maintaining rural roads. For investments like rehabilitation and upgrading the proportions may be higher. In the case of unclassified roads, to which local communities contribute by

offering volunteer labor or other in-kind services, the proportion financed by the road fund may be lower, reflecting the difference between the market wage and the opportunity cost of labor in the countryside.

How should funds be disbursed to each road agency? The road fund must have procedures for disbursing funds to the road agencies. These procedures should be used to strengthen financial discipline. There are three main ways of structuring disbursement procedures. The road fund can either: disburse funds directly to the road agencies on a regular basis and then audit use of the funds ex post, issue approval for the work to be done and then reimburse the road agency after the work has been completed, or pay the contractors directly, but only after certification that the work has been completed according to specifications.

The first method works best when there is good governance, competent road agencies, and a highly decentralized road administration. The road fund finances road agencies if they agree to allow the road fund to review and audit the application of these funds, usually against an approved annual expenditure program. Staff from the road fund visit each road agency to review their internal financial systems (including accounting and related financial systems) to confirm that they are being used correctly and in conformity with the road fund's policies. Technical and financial audits are carried out regularly, and procedural audits are carried out every few years to check on the custody, recording, and use of road fund resources. If a road agency does not comply with the procedures laid down by the road fund, it may have to repay the funds received. Transfund New Zealand operates this type of system.

The second method functions like a line of credit. The road fund first defines the systems and procedures that each road agency must follow. These may involve an annual audit carried out by independent auditors, covering both financial compliance and internal control procedures. Staff from the road fund usually check these procedures on an ad hoc basis and may also carry out field inspections of work financed through the road fund. Each year an expenditure plan is approved, and the road agency goes ahead with implementation. Once work has been completed, the road agency pays the contractor and submits a voucher for reimburse-

---

Box 8.11 Cost-sharing arrangements for maintenance and investment

Maintenance

CANADA. Several provinces share maintenance costs with local authorities. Ontario provides basic funding that can be used to finance maintenance. Cost-sharing is generally 50 percent for urban municipalities. For rural municipalities it is based on ability to pay, with the province generally paying 60 percent of costs. For unincorporated areas, that is, roads managed under local road boards, the province meets two-thirds of basic maintenance costs.

JAPAN. The central government finances half the costs of maintaining directly managed national highways. The remainder is financed by prefectural governments and designated large cities.

NEW ZEALAND. The road fund finances part of the costs of maintaining local authority roads. The proportion financed is equal to $k_1 + k_2 \log (P/LV)$, where P is the current year allocation (thousands of dollars), LV is the three-year average net equalized land value (the local property tax base in millions of dollars), and k_1 and k_2 are constants computed to ensure that the average national proportion is 50 percent. If the calculation results in a proportion that differs by more than 2 percent from the previous year's, the proportion is adjusted by half the difference to be within 2 percent of the indicator. If the calculation results in a difference of less than 2 percent, no change is made unless the trend continues for two consecutive years. The actual proportions in 1996–97 varied from 43 percent to 83 percent (the latter being an offshore island, which is an exception).

FINLAND. FinnRA and local authorities provide funds for maintaining unclassified roads managed by road cooperatives. FinnRA finances the following amount per km: $0.75*L*\$800*(L-0.1*R)*C$, where 0.75 is the maximum proportion of costs to be financed, L is the length of the road, \$800 is the estimated average maintenance cost per km, R is the number of permanent residents living along the road, and C is the maintenance class of the road ($C = 1.50$ for class 1 roads, 1.25 for class 2 roads, 1.00 for class 3 roads, and 0.75 for class 4 roads). In 1990 FinnRA and the municipalities financed 25 percent and 33 percent, respectively, of the maintenance costs of these roads.

Investment

CANADA. Provinces also provide funds for new investment. The basic funding in Ontario can be used for either maintenance or new investment on the same cost-share basis. Supplementary funds are available for specific capital projects and under various special programs. Under the strategic transportation improvement program costs are shared equally among the federal government, the province, and the municipality. For unincorporated areas, that is, roads managed under local road boards, the province meets two-thirds of basic construction costs and finances 100 percent of specific projects.

JAPAN. The central government finances two-thirds of the costs of improving directly managed national highways, 70 percent of the national expressway network, and half the costs of subsidized national highways, main local (prefectural) roads, and main local (municipal) roads.

JORDAN. The Irbid municipality charges part of the costs of road and street improvements to adjoining property owners. It charges 10 percent of the costs of upgrading and widening to landowners on each side of the road, the entire cost of constructing sidewalks, and if the land-take is greater than 25 percent of an individual's holding, they pay compensation only for the excess over 25 percent.

NEW ZEALAND. Funds for investment are provided on the same cost-share basis as for maintenance.

---

ment to the road fund. If the work is done using in-house staff and equipment, the agency submits a claim for bought-out plant and materials together with in-house staff costs. The road fund certifies the payment and then transfers the necessary funds. The mandatory certification prior to payment gives the road fund additional control over the payment process. This system is used by the U.S. Federal Highway Trust Fund.

The third method involves more oversight by the road fund. It starts with the same approved expenditure program, but instead of transferring funds to the road agencies, the road fund pays contractors directly. Payment is made only after certification that the work has been carried out and completed according to specifications. This procedure works best when the work is done under contract, but can also be applied to force account work. Certification is usually done by a firm of local consultants. In practice, the contractor typically submits monthly statements of work completed (unless contracts are very small), and within a fixed period of time the consultant certifies the work. These procedures are widely used in Africa for force account work (Benin), work done by state contractors (Mozambique), and work done by private sector contractors (Zambia).

Operational Questions

How should day-to-day management be organized? The staff of the road fund must organize board meetings, assist in developing and publishing the procedures to be followed by road agencies, manage the road fund's income and expenditures, and keep proper accounts to ensure that the road fund can be audited. Managing income and expenditures proactively requires staff that will collect revenues, manage cash balances, establish withdrawal procedures, oversee the use of funds by the different road agencies, and prevent unauthorized withdrawals from the road fund (that is, prevent raids on the road fund).

Collecting revenues involves drawing up agreements or legal contracts with the collection agencies. They usually include the oil companies, the transport ministry (for license fees and fines for overloading), and the customs department (for transit fees). Some of the revenues may also be collected by nongovernment entities. In all cases a formal contract should be signed with the collecting agency—or at least a written memorandum of understanding. The document should spell out procedures for collecting revenues, how funds are to be deposited into the road fund bank account (or into the consolidated fund for transmittal to the road fund), the information to be supplied to the road fund, and the fees payable to the collecting agency. Road fund staff must track movements in the chargeable base (such as sales of diesel and gasoline and the base price), estimate how much revenue should have been collected, adjust the figures for exemptions and rebates, and then reconcile these figures with the amount actually credited to the road fund during the period concerned.

Day-to-day management of funds involves making projections for revenues, commitments, and disbursements. Based on cash-flow projections, the board will have to decide how to handle short-term borrowing and what to do with any cash surpluses. Since the road fund should maintain only a small cash surplus, the funds should be invested in short-term securities like interest-bearing savings accounts and the overnight money market.

Procedures for withdrawing funds should be made as simple as possible. It is usually sufficient to have each check drawn on the road fund signed by one member of the board and the road fund's senior accountant. There should be alternates in case one of the nominated signatories is unavailable. Road fund staff must also oversee the use of funds by each road agency. This involves specifying how investment and maintenance programs are to be prepared, advising road agencies on how to prepare their road programs, and auditing the results to ensure that the procedures laid down by the board have being correctly applied.

Finally, there is the important job of preventing unauthorized withdrawals. Unfortunately, this is a constant preoccupation for many road funds in countries where governance is weak and ministers and senior civil servants frequently attempt to use money from the road fund to finance other government programs or to otherwise promote their own private interests. Some road funds have developed ingenious ways of discouraging raids, like making judicious leaks to the press, hiding revenues in provincial bank accounts, and even issuing a check, calling an emergency meeting of the board, and then canceling the check. But the best strategy—apart from having a strong board—is to keep a minimum cash balance in the road fund account. Work should be programmed—and charges adjusted—to ensure that revenues and expenditures are closely matched, avoiding the need to hold a large cash balance.

Which financial rules and regulations are needed? The financial rules governing management of the road fund are usually included in the legislation, published as legal regulations in the government gazette, or published by the road fund board. Publishing them as legal regulations in the gazette makes it more difficult for rogue ministers to pressure the board into amending the regulations at short notice to suit their own particular interests. Indeed, some countries even lay down the procedures to be followed when making regulations to prevent undue political interference.

The South African Water Services Act, 1997, requires that the responsible minister, before making regulations under the Act, publish the draft regulations; invite comments from a wide array of interested parties; consider all comments received; on request, report on how specific comments have been taken into account; and submit the draft regulations, together with amendments, to the Regulation Review Committee.

The rules and regulations governing management of the road fund spell out the way in which the road fund operates. The main issues covered in the regulations normally include: the purpose of the road fund—which types of expenditures it can and cannot finance; the terms of reference for the road fund administration; the procedures for nominating and appointing members of the board and for resignation and termination of board membership; the functions of board members and the relationship between the board and the executive secretary or chief executive; the terms of reference for the executive secretary or chief executive and how he/she is to be appointed; the role of the secretariat, including its size, terms of appointment, and other conditions of service; and the powers of the minister relative to the board. Sample regulations are included in annex 5 (for a road fund being set up under new legislation) and annex 6 (for a road fund being set up under existing legislation).

How should the road fund be audited? Once road maintenance is fully funded, the road fund will be handling large sums of money and it is important to ensure that these funds are properly accounted for. That is the purpose of the audit. The audit should normally include: examining the records of third parties responsible for collecting the revenues attributable to the road fund to ensure that all the revenues have been collected and promptly paid into the correct road fund accounts; auditing payments made from the road fund to ensure they are supported by adequate documentation and are in accordance with the purposes allowed in the legislation and supporting legal regulations; verifying that that the work financed from the road fund was carried out according to specifications; auditing the transactions and balances of the bank accounts maintained by the road fund; reviewing the accounting and internal control procedures used by the road fund to determine their adequacy; and reviewing the accounts, files, records, and reports of the road fund to determine their adequacy.

The audit report is normally submitted to the minister of the parent ministry no later than three months after the end of each financial year. The minister then submits the report to parliament. The auditors' report may also be included in the road fund's annual report. The accounts may be audited by independent auditors appointed by the board, by the auditor general's office, or by an independent firm of auditors selected by the auditor general. The audits carried out by the auditor general's office are surprisingly thorough, although this varies by country. Audits by independent auditors can be more variable.[5] The qualifications and integrity of the auditors should therefore be closely monitored.

Key Recommendations and Conclusions

The above discussion leads to the following general conclusions and recommendations:

• Many countries already have road funds, although most are poorly designed and do not deliver a secure and stable flow of funds nor do they strengthen financial discipline. The most problematic "first-generation" road funds are in Eastern Europe and Africa.

• First-generation road funds suffer from systemic problems. The most important involve difficulties collecting revenues, unauthorized withdrawals ("raids") from the road fund, payment for goods and services that were either substandard or never delivered, and poor financial management of accounts.

• Recent discussions between the World Bank and the IMF, together with ongoing discussions with ministries of finance in developing and transition economies, have identified the key principles for making commercially managed road funds viable. First, introduction of the road fund should be part of a wider agenda to commercialize road management. Second, only road-user charges (generally vehicle license fees and a fuel levy) should be paid into the road fund. Third, the arrangement must not abstract revenues from other sectors (additional spending on roads must come from extra payments by road users). Fourth, the road fund should be overseen by a representative public-private board and managed by a strong, independent secretariat. And fifth, independent technical and financial audits must be carried out.

• All new road funds should be set up according to the above conditions, and all existing road funds should be restructured or closed down. Furthermore, national road funds should finance all roads, since road users pay for the use of all roads. The road fund should fully finance the expenditures associated with main roads, while local government and community roads should be financed on a cost-share basis. Subsidiary road funds may be set up to finance urban or rural roads.

• The road fund should have a firm legal basis. If it is set up under existing legislation, a sunset clause should be introduced to determine the date by when basic legislation should be passed, or the road fund be closed down.

• The road fund should be overseen by a representative management board that may be a subcommittee of the main road agency board. Members should be nominated by the constituencies they represent and there should be an independent chairperson. Members should be appointed for three- to four-year terms and should be eligible for reappointment for at least one more term. The director of roads should not be a member of the board but should have observer status.

• The road fund should be managed through a separate administration or through a separate division of the main road agency (the latter arrangement may create a conflict of interest if several road agencies are entitled to receive money from the road fund). The administration should be managed by a chief executive appointed by the board, and the chief executive should appoint all staff. The road fund need not employ a large staff, but they must be qualified in engineering, planning, and finance.

• Revenues should be collected using a simple two-part tariff consisting primarily of a fuel levy, vehicle license fees, a supplementary heavy-vehicle fee, international transit fees, and fines for overloading. The tariff must be designed to ensure that it does not extract revenues from other sectors. Extra spending on roads must be financed through extra payments by road users. When feasible, non-road users should be exempted from paying the diesel levy.

• There should be consistent procedures for raising and lowering the road tariff. The road fund board should either be empowered through the legislation to set the road tariff, or the board should recommend the level to the ministry of finance for inclusion in the annual budget.

• There should be consistent procedures for allocating funds among different road agencies. The proce-

dures should be prepared and published by the road fund board.

• There should be cost-sharing arrangements for local government roads. At the community level these arrangements should permit local contributions to be made in-kind (through volunteer labor, use of farm equipment, and so on).

• The procedures used to disburse funds to the various road agencies entitled to receive money from the road fund should be designed to strengthen financial discipline and guard against weak governance.

• The legislative instrument used to set up the road fund should be supported by published regulations or procedures. Ideally, these should be published as legal regulations in the government gazette.

• The road fund should be subjected to regular technical and financial audits, carried out either by independent auditors, by the auditor general's office, or by auditors appointed by the auditor general.

Notes

1. Road transport operators regularly made this claim during road maintenance workshops in Africa, Latin America, the Middle East, and Asia. Furthermore, a recent survey carried out by the Automobile Association in South Africa established that 88 percent of road users were willing to pay an additional fuel levy of about $0.015 per liter provided the proceeds were spent on national roads.

2. The narrow focus of the South African road fund is unusual—the result of a historical accident. It was set up in 1935 when there were few roads and when the main concern was to develop a national network to connect the provinces. It has remained like that ever since.

3. Georgia specifies the fuel levy as a percentage of the value added on fuel. But it is extremely difficult to estimate the value added, making it difficult to collect the fuel levy.

4. Early statistical work showed a weak positive relationship between earmarking and the proportion of investment devoted to roads—hardly a good recommendation, since nearly all these road funds were established in countries where road maintenance was grossly underfunded. See Eklund (1967).

5. A recent audit report carried out by a private firm of auditors raised no issues about the toll revenues collected by a road agency and deposited into the road fund. But an informal spot check by a visiting toll road company noted that, "The only control [over toll revenue] was tickets torn off a roll and handed to the motorist as a receipt. In many cases the drivers paid and drove off without taking a ticket. In these cases, the toll is practically lost as the payment has not been recorded."

9. Introducing Sound Business Practices

The road sector is big business. Many main road agencies are among the *Fortune* Global 500.[1] The Japan Highway Public Corporation manages assets ($216 billion) roughly equal in value to those of General Motors and Sumitomo Life Insurance, the U.K. Highways Agency ($80 billion) is in the same league as IBM and AT&T, while a relatively small road agency like the Roads Department in South Africa ($7.3 billion) is in the same league as Northwest Airlines and Fuji Electric.[2] On the revenue side, some of the larger road funds and toll road operators also rank among the Global 500. The Japan Road Improvement Special Account has roughly the same turnover ($30 billion per year) as Nippon Steel and Pepsico, while the U.S. Federal Highway Trust Fund ($21 billion per year) and Japan Highway Public Corporation ($17 billion per year) are in the same league as Dow Chemical, Lyonnaise de Eau, and Chibu Electric power.

Public sector road agencies will function more efficiently if they are faced with some form of competition or a surrogate for competition. Competition creates market discipline, which is the primary factor motivating managers to cut waste, improve performance, and allocate resources efficiently. Previous chapters have suggested creating such discipline by introducing an explicit road tariff (to encourage users to demand value for money), linking revenues and expenditures (to create a hard budget constraint), and involving users in road management (to ensure that customers are able to influence the type and volume of road services provided by the road agency). Another complementary option is to unbundle services and contract them out.

The above strategies strengthen market discipline and provide managers with the incentive to operate

efficiently. The corollary is that managers must work within an organization that can respond to market discipline. They need:

- A clear and unambiguous corporate mission.
- A strategy to separate planning and management from the implementation of road works (which may involve contracting out implementation to the private sector).
- Effective ways of contracting work out (generally to the private sector).
- An appropriate number of technically qualified staff.
- A sound management structure.
- Appropriate management information systems.
- Appropriate financial accounting systems.
- Procedures for controlling quality of road works.
- Sufficient autonomy to enable them to manage the road agency efficiently.

Defining the Corporate Mission

The first task is to establish the role of the road agency. Road agencies around the world are increasingly setting down a vision or mission statement, from which they derive their principal operational or statutory objectives. The following are selected examples of such statements:

- FinnRA—"FinnRA is responsible for public roads and they make it possible for road users to travel safely and conveniently."
- National Highways Authority of India—"To meet the Nation's need for the provision and maintenance of the National Highways network to world standards within the strategic policy framework set by the Government of India and thus promote economic well-being and the quality of life of the people."

• South African Roads Department—"To deliver an efficient, reliable and safe national road network; and to manage, maintain and improve the national road network."

• Swedish National Road Administration—"To provide the public at large and the productive sectors in different parts of the country with a satisfactory, safe and environmentally friendly traffic service at the least socio-economic cost."

• Transfund New Zealand—"To allocate resources to achieve a safe and efficient roading system."

• Transit New Zealand—"To operate a safe and efficient state highway system."

• U.K. Highways Agency—"To secure the delivery of an efficient, reliable, safe and environmentally acceptable trunk road network," by, among other things, "being respected for excellence and environmental sensitivity in managing and developing our valuable national road network."

All of these statements express a desire to serve the road user, the environment, and the tax payer by making the road network safer, more reliable, more environmentally acceptable, and more efficient. In other words, they are encouraging the road agency to become more customer-oriented.

Within the context of these vision statements, the parent ministry is increasingly setting a series of clearly specified performance targets for the road agency. For example, the performance targets set for FinnRA during 1994 were:

• FinnRA's activities will result in a reduction of 70 injury accidents and seven fatal accidents (minimum target). The total number of accidents on public roads will be less than 3,900.

• FinnRA will reduce the pollution of ground water caused by road traffic and operations, reduce noise pollution, and improve the roadside milieu.

• The target level of service for the main roads during winter will be achieved during 88 percent of winter days.

• The current condition of the road network will be maintained. The ruts on the main roads will not exceed 20 mm. The maximum length of defective pavements shall be less that 7,500 km and that of structurally poor roads shall not exceed 1,200 km.

• The cost efficiency of FinnRA will improve by 1.0 percent.

• The maximum administrative overhead for FinnRA will be less than $105 million.

• FinnRA's return on investment will be at least 5 percent, and the rotation speed of investment will exceed 1.1 (see below for a definition of these indicators).

The performance targets set for Transit New Zealand and the U.K. Highways Agency are similar.

Several developing and transition economies have mission statements, but few have specific performance targets for their road networks. Instead, they tend to have general targets related to the condition of the road pavement (percentage of the network in good, fair, and poor condition) and to the status of the road agency and its assets. The latter targets normally emerge from the governments' general macroeconomic policies, which they expect the road agency to follow. Although these vary significantly among countries, they typically relate to achieving a proper balance between work done in-house and work done under contract with the private sector; doing as much work as possible using labor-intensive work methods (particularly in Africa and Asia); transforming parastatals—including government plant and equipment pools, quarries, and ferries—into autonomous agencies operating along commercial lines (or privatizing or liquidating them); encouraging private sector interests to construct and operate toll roads under concession agreements or to design, build, finance, and operate existing roads under concession agreements; and reducing staff numbers as part of overall civil service reform.

Although the above targets may be specified in less precise terms, it is still worth attempting to develop both a mission statement and a set of performance targets to guide the road agency's operational program. The targets should be set down in writing and included in an annual business plan or the equivalent. Each subsequent plan should then review the extent to which the targets have been achieved and, if they have not, why not.

Separating Planning and Management from Implementation of Road Works

Most countries are actively trying to separate planning and management of roads from implementation of road works—for two main reasons. First, road agencies have too many conflicting responsibilities. They are typically responsible for planning, managing, and executing road works. Since they are both the customer for, and supplier of, the work they finance, there is an obvious conflict of interest that weakens financial discipline and compromises efforts to control costs and maintain quality. Second, road agencies are usually public monopolies and are thus not subject to much market discipline. As a result the costs of road works are frequently 20 to 30 percent higher than work subject to competition. Some of the most startling evidence on the impact of market discipline comes from New South Wales in Australia. In 1991 their Roads and Traffic Authority decided to start contracting work out to the private sector and to expose their own in-house work to more outside competition. Four years later the costs of in-house work had fallen by approximately 25 percent, while the costs of work done by contractors had fallen by approximately 37 percent.

Countries have tackled this problem in three main ways. They have maintained their integrated structure, but assigned the procurer and producer functions to separate divisions or departments; divided the road agency into separate client and civil works organizations (that is, kept the procurer separate from the producer); or kept the road agency as the procurer and contracted out all producer functions to the private sector.

Norway has chosen the first option—it has kept the road management and production functions within the same overall organizational structure, but has separated these functions at the county level. This structure eliminates some of the conflict of interest but does not really address the cost and quality issues associated with in-house implementation. Norway chose this structure probably because of its more relaxed attitude toward competition. Contractors carry out only 25 percent of periodic maintenance, and all routine maintenance continues to be done in-house. Furthermore, the domestic market for contractors is buoyant, and there is little pressure from the public to contract more work out to the private sector. The Public Road Administration is nevertheless concerned about the costs and quality of its in-house work and is currently implementing a detailed cost accounting system that

will enable the Administration to compare its in-house costs with private contractors on an item-by-item basis (see box 9.5).

New Zealand, Finland, and Sweden have chosen the second option. New Zealand separated road management from production in 1953 when management was transferred to the National Roads Board, while production remained with the Ministry of Works. It stayed like this until 1989, when Transit New Zealand was established to manage the trunk road network and the Land Transport Fund. The production function stayed in the Ministry of Works, but was separated into a civil construction division and a consultancy division, both of which were later corporatized. In 1991 the two divisions were made into separate subsidiary companies and have now been privatized. Finally, in 1996 the funding of roads was transferred to Transfund, and Transit New Zealand became exclusively a road management agency.

Finland and Sweden are still in the process of separating road management from production. The activities have already been relegated to different organizations, and the production units are increasingly having to bid for work in competition with the private sector. In Finland 90 percent of all construction and periodic maintenance and 60 percent of routine maintenance is subject to competitive bidding. In Sweden all periodic maintenance and nearly all routine maintenance are subject to competitive bidding. These changes have increased productivity by about 25 percent. The intention in both countries is to fully commercialize the production functions—perhaps in the form of a limited liability company—in the near future and eventually to privatize them.

Sweden has carried the process of adopting commercial management through to its logical conclusion. It has effectively turned the road agency into a holding company with subsidiaries. The parent company is managed by a board that delegates day-to-day operations to a director general. The headquarters, together with seven regional offices under the director general, handle traffic safety, environmental issues, transport planning, sector and regulatory issues, public road management, economics and finance, personnel, public relations, and administration.

All production units are organized as subsidiaries—called "result units"—which fall under the authority of the director general, but have a high degree of autonomy. They each have a director and an advisory committee. The advisory committee includes no more than eight members (five appointed by the director general) and it elects its own chairperson. The director is appointed by the director general following consultation with the advisory committee. The production unit, which is required to earn at least a 15 percent rate of return on its equity, currently wins about 60 percent of its work through competitive bidding. The long-term goal is to turn all of the result units into autonomous legal entities competing with the private sector on an equal basis.

The United Kingdom and many other countries, particularly developing and transition economies, have chosen the third option. Until the early 1980s the trunk road network in the United Kingdom was managed by the Department of Transport, virtually all design and supervision of motorways was undertaken in-house, and maintenance was carried out under standard agency agreements with local authorities who relied on their own direct labor. These arrangements were restructured in four broad phases. First, the in-house design and supervision of trunk roads was privatized by inviting bids from consultants to take over the staff and ongoing work programs. Second, when the metropolitan councils were dissolved in the mid-1980s, the maintenance that they previously carried out on an agency basis was competitively tendered. Third, in 1994 the planning and management of roads was transferred from the Department of Transport to a new Highways Agency that operated along commercial lines. Finally, the trunk road network has been divided into 24 zones, and all maintenance work is now being competitively tendered. Work can be contracted out to private sector consortia, local authority consortia, or joint private sector/local authority entities.

Identifying Effective Ways to Contract Out

When work is contracted out, the road agency normally has a choice between the form of the contract and the type of specifications to be used. With regard to the form of the contract, there is a choice among lump sum contracts, in which payment is based on a single price

for the total work; admeasure contracts, in which payment is based on the quantity of completed work, valued at tendered rates in a bill of quantities; cost-reimbursable contracts, in which payment is based on actual costs (this form requires "open book" accounting) plus an agreed fee to cover overheads and profit; and target-cost contracts, in which payment is based on actual costs, plus a fee, together with an additional incentive payment related to any savings beyond the initial target costs.

The main difference among these forms relates to the way the contract allocates risk between the contractor and the client. In lump-sum and admeasure contracts—which are essentially price-based contracts—the contractor bears much of the risk and has to price the tender accordingly. When risks are high, bid prices must be correspondingly high, or else contractors will be reluctant to bid at all. Cost-reimbursable and target-cost contracts, on the other hand, are cost-based contracts for which the client bears the main risks. But the target cost contract also includes an incentive for the contractor to work efficiently and minimize costs. Cost-based contracts are staff-intensive, require good cost accounts, and work only when there is good governance.[3] Most road agencies in developing and transition economies thus favor price-based contracts, particularly admeasure contracts.

Procedural specifications, in which the client defines what work is to be carried out (hence often referred to as cook-book specifications), are traditionally used for roads—with good reason. Procedural specifications are relatively easy to identify and measure, particularly for new construction. But they require a lot of supervision, and since the contractor cannot easily change the design, work methods, or materials, there are few incentives to encourage contractor innovation.

In recent years these problems have led to a gradual move in some countries from procedural (or method) specifications toward functional (or end-product) specifications, in which the client determines the desired level of service required in terms of clearly defined functional or performance characteristics (for example, specifying pavement performance in terms of roughness, rutting, surface friction, and so on). Such specifications help to minimize the amount of supervision required, since it is necessary to test only the performance of the facility, rather than each item contributing to that performance. Functional specifications also encourage contractors to find the best way of meeting the performance requirement (for example, by maximizing use of their own particular skills, equipment, and materials). Experience with performance contracts has been encouraging. The Road Transport Authority in New South Wales, Australia, let a 10-year performance contract in 1996 that has reduced maintenance costs by 48 percent over their own 1991 in-house costs. The main difficulty with such contracts is the need to define and describe the functional or performance specifications.

The scope of works covered by the typical road contract has also begun to change. For example, periodic and routine maintenance are usually contracted separately, often using different standard contract documents. But some countries, including Algeria and Brazil, are now letting combined contracts for execution of routine and minor periodic maintenance works on specific road sections (average length 244 km in Brazil). Other countries, including Canada (British Columbia), Chile, Colombia, Malaysia, the United Kingdom, and Uruguay, are combining all maintenance works on specific routes, or within entire geographic areas, in comprehensive maintenance contracts that run for several years. The contracts are currently running for five years in British Columbia; three and five years for unpaved and paved roads, respectively, in Chile; two years in Malaysia; three to five years in the United Kingdom;[4] and four years in Uruguay. Sweden is using contracts that run for three to six years. The United Kingdom is now also letting 30-year performance contracts for roads requiring major rehabilitation or new investment. Contract payments are either indexed to traffic flows with the contractor being paid through shadow tolls or are based on lane availability. Finland has recently let a similar contract for 15 years.[5]

Contractors in these countries prefer contracts that run for at least five years, since this amount of time provides sufficient incentive to invest in specialized equipment. These innovations bring considerable cost savings, but require well-developed contractors and a highly professional road agency to make them work.

A number of incentives are also being offered to contractors—in addition to those offered in target cost contracts—to encourage them to be more innovative. For example, some contracts require the contractor to design, construct, and guarantee a road pavement for a specified period of time. Such contracts are being used in some parts of the United States and are being piloted in South Africa.

The United Kingdom is proposing an additional performance incentive by extending the contractor's guarantee period from the usual 12 months to 36 months. The United Kingdom, especially plagued by congestion, has also been trying to provide the contractor with an incentive to complete road works as quickly as possible. This incentive is mainly used on high-volume roads, where site works are hazardous and seriously disrupt traffic. In regular price-based contracts liquidated damages can be imposed for work completed late, thus giving the contractor an incentive to complete work within the specified contract period. But damages do not provide any incentive to complete road works early. The need for such incentives led to the development of lane rental contracts. These contracts provide an explicit financial incentive to encourage the contractor to complete the work as fast as possible (box 9.1). The design, build, finance, and operate contracts being let in the United Kingdom and Finland include lane closure penalties—graduated by time of day—to encourage early completion.[6]

One of the major constraints hampering contracting out—even for relatively simple road works—is the underdeveloped nature of the local consulting and construction industries in many developing and transition economies. The road agency cannot invite competitive bids unless the country already has consultants and contractors with road work experience. Furthermore, the road agency cannot be expected to prepare bid documents, award contracts, and supervise implementation of civil works using staff accustomed only to doing work in-house. Staff in the road agency must know something about the preparation of bid documents, contracting procedures, contract law, and arbitration procedures before the road agency can effectively contract road works out to private firms and hire consultants to design and supervise implementa-

---

Box 9.1 Incentives for early completion: lane rental contracts in the United Kingdom

Lane rental contracts were introduced in the United Kingdom by the Department of Transport in 1984 on major schemes to speed up maintenance works and reduce delays. According to the basic contract arrangement a bonus is paid if the contractor finishes the works before the contract completion date, but a charge (at the same rate as the bonus) is imposed if the contractor is late. The contract replaces liquidated damages with a daily charge for late completion that is related to the costs imposed on road users. Bonuses paid on individual schemes have ranged from $8,000 to $1.6 million. Both bonuses and charges are based on an assessment of the economic costs to road users of delays. Two variant systems, continuous site rental and lane-by-lane rental, were established one year after the concept was first introduced.

• Bonus/rental charge. The contractor tenders a price for the work and time of completion, and receives a bonus or pays a charge according to the number of days that work is completed ahead of or after the contract period.

• Continuous site rental. The contractor is charged a daily rental rate for each day that there is possession of the site.

• Lane-by-lane rental. The contractor is charged according to the number of lanes occupied.

In setting the appropriate bonuses and charges, the Department of Transport considered that it would be too costly to pay bonuses at the full daily delay rate and that a lower rate would still provide the contractor with adequate incentives. They also took into account the probability that charges based on the full daily rate might not be recoverable or might deter companies from tendering. It was therefore decided that the bonus should normally be set to cover 50 percent of the daily delay cost plus 100 percent of the daily site supervision cost. Later, the Department limited the daily bonus/charge rates to a minimum of $3,000 and a maximum of $40,000.

Between 1984 and 1989 the Department of Transport let about 100 lane rental contracts worth approximately $400 million. It is estimated that these contracts saved more than 2,400 days of lane closures, representing an economic savings of about $80 million, at an additional cost of $13 million paid in bonuses. Although there is some debate about the actual size of the benefits, a detailed comparison of contracts in 1987-88 estimated that the average rate of spending on a lane rental contracts per week was 81 percent higher than that for a conventional contract, confirming that lane rental contracts have quickened work substantially and reduced traffic delays.

Source: U.K. National Audit Office, 1991.

---

tion. Most efforts to promote competition therefore must be accompanied by parallel efforts to develop the local construction and consulting industries (box 9.2).

Staffing Requirements

Once the road agency has developed a mission statement and defined its key performance targets, it can turn its attention to the number and type of staff needed to run its operational program. If the emphasis is on contracting out, the operational program could be handled by fewer staff, but they will need different qualifications. The small regular staff could be supported by small-scale local contractors for most routine maintenance work, small-scale contractors for the rehabilitation and periodic maintenance of gravel roads, and medium- and large-scale contractors for the patching, periodic maintenance, and rehabilitation of paved roads (table 9.1).

The figures in table 9.1 suggest that a fairly efficient road agency, which contracts most work out to the pri-

vate sector, should be able to plan and manage the network with five or less staff members per 100 km. On networks with heavy traffic, where additional in-house staff may be involved in traffic control and driver information activities, the number per 100 km may be closer to 10. A main road network of about 15,000 km (10,000 km paved) with moderate traffic and organized into 10 maintenance districts, may thus require about 300 to 750 regular staff to plan and manage the network (about 50 managers and the remainder working as engineers, technicians, administrators, and other support staff). Between a quarter and a third might work in headquarters support services, while the rest would be located in regional offices. Subcontractors would do the balance of the work, and they might in turn employ an additional 1,500 staff if the work was done using capital-intensive techniques or 7,500 staff if the contractors were using labor-intensive techniques. By these standards most road agencies—which typically employ several thousand workers—are overstaffed, primarily because they have not yet separated

Box 9.2 Developing domestic contractors for road maintenance

A number of initiatives have been taken to develop the capacity of local contractors. They include providing preparatory and hands-on training, providing access to plant and equipment, helping road agencies to acquire the skills needed to supervise contracts, simplifying government procurement procedures, and setting up, adapting, or strengthening permanent education and training institutions for road specialists.

PREPARATORY TRAINING. Seminars have been organized in transition economies (such as Estonia, Latvia, Lithuania, Poland, and Vietnam) to introduce consultants, contractors, and civil servants to competitive bidding, cost control, and contract management. Similar seminars have been organized in Africa to teach contractors how to manage small civil works contracts. The most comprehensive training program was given in Tanzania for administrative managers, engineers, site superintendents, and technicians. Owners and managers of the firms were asked to participate in the training so that they could understand what was being taught to their staff. The most creative scheme, in Malagasy, used multimedia techniques to get the message across.

HANDS-ON TRAINING. Potential contractors have been permitted to work on small projects to gain practical contract experience. Hands-on training covering labor-based construction techniques has been used to develop small firms for over two decades in Latin America, particularly the *micro-empresas associativas* in Colombia and the Dominican Republic. Similarly, in Guinea Bissau the International Labor Organization has organized 3-km training sections for labor-based rehabilitation of feeder roads. In Kenya contractors have been trained to bid for road rehabilitation works. First, unit prices were fixed by the road agency. Then, contractors were allowed to bid with the same rates but with a plus or minus factor. Now they have to compute their unit prices themselves. Current contracts amount to about $500,000 each.

AVAILABILITY OF PLANT AND EQUIPMENT. These initiatives help contractors gain better access to plant and equipment. In Uganda rented equipment belonging to the Ministry of Works was made available to contractors, but the amount

was not sufficient. Contractors therefore decided to buy additional equipment and share it through a pool. Ministries of works in several African countries are considering renting equipment to contractors, while several donor-financed projects are providing contractors with foreign exchange that enables them to buy equipment and spares.

CONTRACT SUPERVISION. Most road agencies have limited capacity to supervise contracts, and several initiatives are under way to strengthen this capacity. Many African countries are building or strengthening control units in the road agency to adequately supervise contracts. In each case foreign experts are involved in compiling sample documents for preparation, procurement, and supervision; staffing the unit during the initial years; and training civil servants in this new activity.

SIMPLIFYING PROCUREMENT PROCEDURES. Simplifying procurement procedures is essential for doing more work under contract and developing the local construction industry. In Ghana a comprehensive review of contract conditions for international and local competitive bidding and LCB has been carried out, and proposals for changes have been prepared and accepted. New conditions are being implemented. If they had been implemented earlier, some specific clauses, such as provisions for compensation for delayed payments, might have prevented some contractors from going bankrupt, although there is no substitute for prompt payment. Other countries have also decided to reshape and simplify the regulations for procurement and contract administration to make them easier for contractors.

SETTING UP PERMANENT TRAINING INSTITUTIONS. In most countries the training center of the ministry of works was the only institution in charge of educating road specialists, and training was often tied to the implementation of foreign-funded projects. Training must be funded on a permanent basis, it must be open to contractors, and the curricula should include contract management. Institutes in former centrally planned economies are in danger of failing because of lack of funding and inappropriate privatization arrangements. It is important to keep these institutions alive and to extend their curricula to contract management and cost control.

Source: Prepared by J-M Lantran for this study.

planning and management from implementation of road works.

Any reduction in the size of the road agency must be accompanied by an improvement in terms and condi-

tions of employment, particularly for older staff with experience, the CEO, and directors. In general, current salaries for engineers and technicians must be more than doubled to make them competitive with private sector

Table 9.1 Road agency staff involved in planning and managing road works in selected countries

Road agency	Length of network (km)	Number of staff	Number of staff per 100 km
Transit New Zealand	10,500	189	1.8
Finnish National Road Administration	78,000	1,500	1.9
Swedish National Road Administration	97,908	2,000	2.0
South African Roads Department[a]	6,133	140	2.3
Ghana Highway Authority	14,100	688	4.9
INVIAS, Colombia	13,408	920	6.9
Korea Bureau of Public Roads	12,053	1,450	12.0
U.K. Highways Agency	10,500	1,600	15.2

a. Staff numbers projected as of end-March 1998.

salaries, as must the salaries of the CEO and directors. Fringe benefits must also be improved. Unfortunately, very few road agencies have managed to better terms and conditions of employment. Even FinnRA and the U.K. Highways Agency are still classified as government departments and employ all staff under civil service conditions, although often with some flexibility within salary ranges. Some road agencies have nevertheless managed to become autonomous (New Zealand), semi-autonomous (Sierra Leone), or have formed into nonprofit joint stock companies (Latvia), enabling them to pay market-based wages and operate in a fully commercial manner.

Restructuring and downsizing are likely to create redundancies. This is normally handled by tackling the problem in stages. The usual first step is to offer older staff incentives to retire early, although this generally affects staff numbers only modestly. The second step is to identify all the commercial activities within the road agency and to move them into units that can then be spun off as separate commercial enterprises (for example, traffic data, civil works design, manufacture of road signs, materials testing). The road agency usually helps the staff to set up these enterprises under what is effectively a management-staff buyout.

Sometimes, as with civil works design, the road agency is able to invite bids for design work from private sector consultants. These contracts include a requirement to take on all, or most, of the in-house staff. The United Kingdom managed to reduce the number of engineers and technicians working on design and supervision of road works by negotiating their transfer to private sector consultants in conjunction with the transfer of agreed design and supervision work for a specified period of time.

Similarly in-house implementation of civil works is often restructured into one or more profit centers—which are then prepared for privatization or for operation as state construction enterprises. Finland, Namibia, and Sweden are in the process of doing this. The road agency may also help in-house laborers to convert into small-scale contractors. Assistance normally takes the form of training, providing credit to purchase standard sets of equipment, and offering of initial trial contracts. The Ghana Highway Authority managed to reduce their staff from 8,400 to 4,700 primarily by converting them into petty contractors. Moving staff into the local construction industry tends to be easier if, along with restructuring the road agency, a road fund is established. A fully funded road maintenance program will usually create more than enough new jobs for former road agency staff. Finally, if all else fails, the road agency must offer exit packages to redundant staff.

A smaller road agency must also address its skill-mix requirements. A restructured road agency will be more commercial and more of a planner, facilitator, and paymaster. It will thus need more managers, more staff with financial backgrounds, and more engineers with experience in contract management, contract law, and arbitration procedures. If the road agency intends to promote labor-based work methods, it will also need staff who know when such techniques are suitable and who can train small-scale contractors to do such work. These changes require new personnel policies and revised training programs. They also call for a new look at technical assistance programs. With a clear mission and competent, well-paid staff, developing and transition economies should not have to plan for long-term expatriates in line management positions. Technical assistance can instead be refocused to meet clearly identified skill needs.

Management Structure

Some main road agencies have already been restructured to create a more commercially oriented management structure (figure 9.1). Several road agencies have established the post of CEO and created a new layer of line managers appointed at roughly the same level as the former director of roads. In the United Kingdom the CEO is appointed on a performance contract and his or her remuneration is partly related to the delivery of the agreed road program. The agencies tend to have five or six line managers (directors) who are responsible for functions like administration, road development, road network management and maintenance, design and environment, and finance. Most also have small units that handle public relations and internal audits.

Such restructuring has also addressed the issues of shared services within the usual ministerial structure and confused reporting arrangements. The agencies have been given their own support services (administration and accounting) instead of having to share these with other parts of a larger ministry of works. Reporting arrangements have also been made more direct. Regional managers report directly to the CEO—instead of to the permanent secretary, as in some countries—and the CEO reports through the board directly to the parent ministry.

The staffing structure within each of the departments has also been simplified and made more relevant to the restructured departmental functions. A traditional road agency typically has a large number of job grades, which are rarely related to the size of the division or department and the needs of the task. There is also too much layering. Restructuring therefore tends to start by preparing new job descriptions based on the redefined functional needs, designing a new organizational structure for each division and department based on the new job descriptions, reducing layering by grouping staff into a limited number of job categories (senior management, middle management, and operating staff), introducing a new reward and career system, and reviewing and revising the disciplinary

Figure 9.1 Proposal management structure for a main roads agency

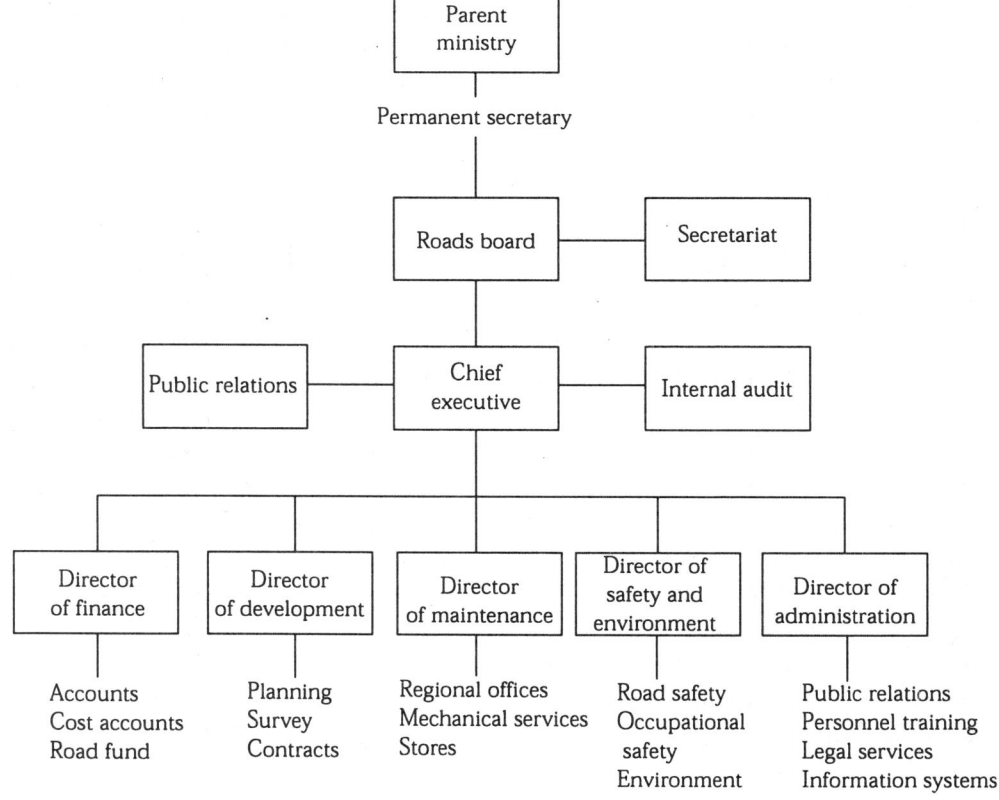

code. The reward system now often includes payment of performance-related bonuses (as in Finland) and wage bonuses tied to annual performance evaluated against agreed road network goals (as in Korea).

The regional structure of the road agency is also important, particularly in large countries where centralized agencies tend to be too remote from their customers. Several countries are attempting to address this issue by decentralizing operations to regional offices. Under such arrangements most planning and execution is done at the regional level, while the headquarters staff coordinate regional programs, operate the management information system, and provide other central support services. More than two-thirds of the staff may thus be placed in regional offices (for example, in New Zealand more than 65 percent of staff work in regional offices).

The U.K. Highways Agency is carrying the concept of decentralization one step further. It plans to create network managers who will be fully responsible for managing part of the road network. The network manager will buy in the services required to meet agreed targets and the headquarters staff will basically become a service unit serving the needs of the network managers. FinnRA and the Swedish National Road Administration are organizing their staffing along similar lines.

Management Information Systems

Management cannot plan, deploy, and control resources without essential information. The road agency's parent ministry and constituents likewise need such information to judge whether the road agency is using resources efficiently and providing road users with value-for-money. The management information system used by the average road agency consists of a set of established and documented procedures that generate and evaluate alternative ways of operating, maintaining, improving, and extending the road network. It will generally show the condition of the road network and its use (traffic volumes and loading), and can be used to explore the impact of management interventions on current and future service levels. It can also be used to generate information on the physical and financial performance of the road network (see table 9.2 for the indicators suggested by the OECD Scientific Expert Group IR7).

The management information system provides a framework for making decisions on a number of issues usually handled by different divisions within the road agency. They include decisions on:
• Carrying out routine and periodic maintenance of gravel roads, paved roads, and bridges.
• Rehabilitating pavements and bridges.
• Upgrading gravel roads to paved standard.
• Improving the geometric characteristics, or capacity, of roads.
• Setting charges for the use of roads and bridges.

Each of the above activities are interdependent with regard to the road agency's budget constraint. Resources have to be allocated among competing programs to optimize expenditures, and user charges must be set to generate the resources required to finance them. The management information system should thus provide the basis for allocating resources to achieve the best overall road conditions.

The management information system will usually comprise data collection, storage, and analysis; estimated traffic, predicted future road conditions; and the impact of alternative management strategies. The system should not be too complicated. Developing and transition economies are littered with the relics of failed management information systems that were poorly designed, emphasized system over management, and were overly complicated. The key guiding principle is that systems should be affordable, suit the decision-making needs of the road agency, be compatible with the scarce human resources needed to operate them, and be capable of being incrementally upgraded when resources permit. The initial system should be simple and unpretentious and should focus on monitoring the condition of the road network. There is no substitute for a monitoring system that includes regular visual inspection of the road network to raise awareness of maintenance needs.

Most road agencies generally begin with a centralized, manual management information system with four modules: traffic information (classified counts with some axle-weight data), a survey database (periodic, visual road condition survey data), road planning (upgrading and new roads), and maintenance man-

Table 9.2 Performance indicators suggested by the OECD Scientific Expert Group

Measure	Parent ministry	Road agency	Road user
Accessibility, mobility	*Primary:* Highway Capacity Manual level of service; average road-user cost (car and truck) *Secondary:* composite access index; total transport cost per GNP	*Primary:* expenditure for maintenance and operation per vehicle km and equivalent standard axles (ditto by functional class); travel time and its variability *Secondary:* quality of information to road users (from audit)	*Primary:* level of satisfaction regarding travel time and its reliability; quality of road user information (both by user group, from market surveys)[a] *Secondary:* hours of congestion delay
Safety	*Primary:* Accident risk: fatalities and injury; accidents per vehicle-km (and the number of fatalities and injured); existence of national traffic safety program/plan; percentage of accidents involving drunk driver	*Primary:* existence of method to assess results of safety programs (yes/no); percentage of traffic speeding (weighted) *Secondary:* percentage of roads not meeting minimum design standards; exposure of pedestrian and cyclists to vehicle traffic	*Primary:* unprotected road user risk *Secondary:* time from alert to treatment (medivac); percentage of population that considers traffic injuries a public health problem
Environment	*Primary:* existence of air quality standards *Secondary:* cumulative land area taken by roads (percent); new land area taken for road use; existence of inspection/maintenance programs for vehicular emissions	*Primary:* environmental policy/program (yes/no); use of de-icing agents; emissions per capita for CO_2, NO_X, VOCs, particulate matter *Secondary:* pollutants in road run-off	*Primary:* percentage of population exposed to noise level greater than 65 decibles *Secondary:* percentage of population exposed to emission levels violating air quality standards
Equity	*Secondary:* regional distribution of roads; laws for mobility limited (yes/no)	*Secondary:* surplus (deficit) of road expenditures relative to road-user charges collected by region	*Secondary:* travel cost, travel time by user group; accident risk by user group
Community	*Secondary:* processes for public participation and procedures to reconsider prior decisions	*Primary:* processes in place for market research and customer feedback (yes/no)	*Primary:* satisfaction with the number and types of feedback mechanisms
Program development	*Primary:* long-term programs for construction, maintenance and operations (yes/no) *Secondary:* benefit-cost analysis of the adopted road programs; projected level of congestion	*Primary:* management system for distribution of all resources (yes/no); benefit-cost analysis of the (proposed) road program *Secondary:* quality management/ audit program (yes/no)	*Primary:* satisfaction with the programs development process
Program delivery	*Primary:* sufficiency of maintenance funding *Secondary:* degree of completion of the long-term road programs	*Primary:* forecast values of road costs versus the actual costs; cost of operations/ lane-km; overhead percentage; percentage of construction materials recycled *Secondary:* number of staff/ lane-km; percentage of work done by direct labor	*Primary:* satisfaction with programs delivery; road administration costs and user delay costs associated with operations and maintenance
Program performance	*Primary:* value of assets (trend); ex-post value of benefit-cost analysis *Secondary:* trends in road budget by road programs (construction, maintenance, operations); return on assets; total road expenditures per GNP	*Primary:* roughness (by functional class); bearing capacity (by functional class); percentage of defective bridges; percentage load posted bridges; deck area; congested road-km; incidence of truck overloading *Secondary:* existence of management system for road furniture	*Primary:* surface condition; satisfaction with road condition *Secondary:* rest areas/ 100 km; percentage of main roads lighted; quality of traffic conditions during winter; user information system (yes/no)

a. Level of satisfaction regarding travel time and its reliability and quality of road user information are combined into one indicator obtained from the same road user survey.

agement (using engineering judgment and standard unit costs). Despite its simplicity, some road agencies, as presently staffed, will not even be able to manage this level of sophistication. For a 15,000 km road network (10,000 km paved), the system would require about two to three traffic count teams, two road inspection teams, and at least one engineer and one technician to operate the system. Traffic counting and inspection could be done by consultants. The engineer and technician should ideally be in-house staff.

The next level of sophistication is probably to computerize the system, but to do so using simple analytical tools and in a way so that it is accessible to regional engineers. The system in Pakistan has currently reached this stage of development (box 9.3). Scoring sheets are used for each section of road, and key features, like cuts, fills, sign posts, and other maintenance features, are recorded in diagrammatic form. The scoring sheets are then used to work out quantities and to prepare estimates that eventually form part of the contract documents. The sheets are also used for scheduling work and, finally, for checking on what work has been done and its quality.

Simple systems can be incrementally upgraded to expand the database and increase the sophistication of the analytical methods used to manipulate the data. The survey database might be extended to include surface roughness and pavement strength, and the pavement management system might be strengthened by basing it on an analytical model, like the World Bank's Highway Design and Maintenance model (version III; version IV is currently being tested). But this takes more resources, requires continuity among the staff operating the system, and should be attempted only when there are sufficient trained staff and other resources. Further sophistication can then follow, perhaps along the lines being pursued in Indonesia, although that level of sophistication lies well in the future for most developing and transition countries (box 9.4).

Financial Accounting Systems

The road agency's financial accounting system should be designed to complement and support the management information system. It should present a clear pic-

ture of the road agency's overall financial health and be able to produce the financial data needed to plan expenditures, compare alternative strategies, monitor implementation, and account for the way funds are used. Standard government accounts, which focus almost exclusively on cash expenditures, cannot do this. A number of road agencies, notably in parts of Australia, Finland, New Zealand, Pakistan, Romania, Sierra Leone, and the United Kingdom, are therefore restructuring their accounting systems along commercial lines to provide a better basis for making informed management decisions. They are generally moving toward regular commercial accounting systems, which include a standard income statement, a balance sheet, and a sources and application of funds statement. Both Transit New Zealand and the U.K. Highways Agency are currently valuing their assets with a view to eventually earning a specified rate of return on capital.

Many of the benefits of commercial accounting can be achieved with relatively simple reforms that do not necessarily involve preparing full commercial accounts. The most important reforms include: preparing an income statement that matches revenues and expenditures; accounting for all the assets owned directly by the road agency (that is, excluding the capital invested in roads); recording, in a simple and transparent fashion, the financial condition of the road network; and producing better information on actual costs to support the above road management systems.

FinnRA and Transit New Zealand currently produce some of the best road agency accounts, and the U.K. Highways Agency is not far behind. This has many advantages, including making it easier to specify performance targets to measure the road agency's overall performance. For example, in Finland the provincial road administrations—there are nine of them, each operating as a "profit center" of FinnRA—are required to meet certain specified financial targets, including the return on investment (operating result/investment), the operating result (operating result/operation revenue), the rotation speed of investment (operation revenue/investment), and increased value per person (gross margin + salaries and wages + rents, all divided by the number of personnel).[7] These performance indicators are then aggregated to become performance indicators for the administration as a whole.

Box 9.3 Simple road management systems: the case of Pakistan

Pakistan's highway network has suffered from a combination of deferred maintenance and rapidly increasing traffic levels. To help remedy this, the National Highway Authority developed a simple Maintenance Intervention Level System to identify maintenance and rehabilitation needs along the entire National Highway System. This system produces numerical ratings that allow engineers to determine how much maintenance a roadway needs and which roads require reconstruction or extensive rehabilitation.

The system works by taking simple, direct measurements of the roadway and its environs. The network is divided into contiguous 5 km sections and measurements are taken along one representative sample km in each section. Eighteen separate factors are measured to determine the maintenance needs of each section. They are grouped into conditions that may be improved by maintenance (such as roughness) and factors that influence how much maintenance is needed (such as climate, terrain, and traffic volume).

Each measurement is assigned a severity score and the scores are combined to obtain a total measure indicating the level of maintenance needed by each section. Scores are forwarded to maintenance field offices where they are used to estimate the amount of materials needed to correct roadway deficiencies. Costs are also estimated, and this information is used to define priorities and prepare a finalized work program that accords with the maintenance budget.

Information collected includes: road type, road width, pavement type and roughness, number of potholes, length and depth of rutting, pavement cracking, axle loading, pavement and subgrade strength, rainfall and available drainage, and edge step and erosion. Rutting measurements are made by laying an aluminum bar across the pavement and sliding a wedge into the space between the bar and the rutted road. The wedge is divided into four colored sections, and technicians record depths according to the color of the wedge. The simplicity of the measurement system means that data collection teams get consistent results. Similar measurement systems are used for cracking, edge step, edge erosion, and subgrade strength.

Severity scores have been developed for each of the 18 factors. These are added together for each 5 km section of road to yield a final intervention score. This score indicates the overall level of maintenance needed on the roadway segment and is plotted on an intervention map. Intervention scores are categorized into one of the following ranges:

- Greater than 70: Requires reconstruction/rehabilitation. Only safety-related maintenance should be carried out.
- 60–69: Road pavement requires improvement. Major periodic overlays should be programmed. A road segment with this rating needs localized periodic maintenance.
- 40–49: Requires routine maintenance. Should be given preferential status over other routine maintenance needs.
- Less than 40: Requires routine maintenance. Should be given attention only as budgets permit.

The Maintenance Intervention Level score has been found to be fairly accurate in indicating the quantity of materials needed to rectify any deficiencies. Some of these relationships, using scores adopted by the National Highway Authority, are shown below.

Erosion from edge-patching work:
Score: 7 up to 125 sq. m edge patching/km
9 up to 250 sq. m edge patching/km
11 up to 325 sq. m edge patching/km
13 up to 500 sq. m edge patching/km

Potholes—carriageway repairs:
Score: 0 expect possible 10 sq. m /km/year
10 expect up to 200 sq. m /km/year
13 expect up to 500 sq. m /km/year
15 expect up to 1000 sq. m /km (7.3 m wide)

Wheel track rutting depth:
Score: 4 = surface treatment
5 = asphalt concrete overlay

Wheel track cracking:
Score: 0–6 = surface treatment
7–8 = asphalt concrete overlay

Both wheel track and center line cracking:
Score: 0–6 = double surface treatment
7–8 = asphalt concrete overlay

After two years of experience with the Maintenance Intervention Level, the National Highway Authority has re-evaluated its intervention scores and assessment procedures. The updated version of the Maintenance Intervention Level uses the same eighteen factors, but categorizes them into four groups: economic factors, pavement condition factors, environmental factors, and road classifications. These categories, combined with revised intervention level scoring techniques, allows NHA to define better necessary maintenance levels. It also permits the National Highway Authority to investigate the effect of applying different maintenance strategies and funding priorities.

Source: Prepared by Robert Butler for this study.

Few developing and transition economies have made similar progress in this direction, although the National Highway Authority in Pakistan has recently started to produce commercial accounts, the National Administration of Roads in Romania has been producing them since 1991, and the Sierra Leone Roads

Box 9.4 The integrated road management system in Indonesia

The current version of the integrated road management system prepares a rolling five-year expenditure plan and detailed one- and three-year programs for all periodic, routine, and holding maintenance and betterment works on the 51,783 km national and provincial road networks. The physical needs, costs, and quantities, and the pretender documents are all prepared automatically, with some allowance for manual interaction to fine-tune the program within specified guidelines. Ongoing and committed projects retain their program status, and the system can now prioritize widening projects in the betterment program on the same economic basis as the entire program.

Project identification and program allocation are based on standardized engineering design and are selected and prioritized to optimize economic benefits at the current official discount rate of 15 percent per year. Detailed allocations, contract packaging, and bidding documents are prepared for the first year of the plan (as well as for commencing multiyear projects, such as betterment), and the system can account for other expenditure programs prepared outside of the integrated road management system. When budgeting constraints are applied, typically to redistribute projects evenly across a three- to five-year period. They are applied according to the incremental economic benefit gained from not deferring a project for a year.

The integrated road management system currently comprises a central database and five application modules, each of which completes a phase of program and project preparation for all roadworks on existing national and provincial roads. The modules are:
• *Planning:* Segments the network into design-level homogeneous sublinks and identifies, designs, and evaluates treatment options for a six-year period to develop a five-year expenditure plan.
• *Programming:* Produces a five-year expenditure plan and a detailed first-year roadworks program for a network, applies any budgetary constraints, facilitates manual refinement of program, and groups projects into pretender packages.
• *Road design:* Performs prebid design of works and estimation of quantities and costs, and prepares bid documents.
• *Economic review:* Makes formal economic evaluation of each contract package to meet requirements of funding agencies.

• *Budgeting:* Prepares budget reports from the work program, assigning estimated costs to relevant budget categories.

A construction implementation module is still being developed. This will manage contract procurement, provide for detailed engineering and design review under contract, monitor and manage routine maintenance, and assist in contract supervision and quality assurance.

The central database is the core resource of the integrated road management system, storing all of the data that the application modules need to operate and the output from those modules. The database also provides the basis for preparing road network statistics and management reports. Similar but smaller databases reside in each province. These contain all road, traffic, and program data collected for the province. The database is now structured in seven data groups:
• *Administrative data:* identifies the regional, province, Wilayah, and Seksi jurisdictions.
• *Road data:* inventory attributes (link description, location reference, road description, road geometry); pavement strength (structure, CBR, BB, deflections, FWD deflections); road condition (roughness, distress, minor structures, and sideworks condition); materials sources (quarry, soil cement zone); and project data (committed and ongoing projects).
• *Bridge data:* inventory and condition data for bridges longer than 6 m (maintained in the bridge management system).
• *Traffic data:* traffic volume, classification, and loading data.
• *Budget data:* annual budgets and five-year plan budgets for national and provincial roads.
• *Cost data:* unit road works costs and road user cost parameters.
• *History data:* historical record of date, type, and costs of works performed on the network.

The system can develop an economically optimal annual works program and five-year plan, including preliminary engineering design, contract packaging, and document preparation for more than 6,000 sublinks and 200 contracts annually. With full implementation in the central road agency and most provincial roads departments, the integrated road management system covers more than 95 percent of the national and provincial network and is among the most advanced systems in the world.

Source: Prepared by W.D.O. Paterson for this study.

Authority has been producing them since 1992. Road agencies generally start by producing an income statement and a statement of affairs. The income statement records the agency's income (for example, income from

the road fund, proceeds from the sale of contract documents, government grants) together with the expenditures associated with operating and maintaining the road network. The statement of affairs is a modest document that simply lists the fixed assets owned by the road agency (vehicles, plant and equipment, and office equipment), money owed to the road agency (debtors), cash in hand, and money that the road agency owes to others (creditors). These reforms represent little more than better bookkeeping arrangements. The next step is to turn the statement of affairs into a regular balance sheet and to add a cash flow statement, which is a simplified sources and application of funds statement.

These financial reforms can have a major impact on managerial behavior. They provide managers with a better record of what is happening to the business; motivate them to locate all of their assets,[8] assign them a value and record that value; encourage a culture of managing assets; and take a first step toward fully costing the overhead and administrative expenses of operating and maintaining the road network. Financial reform is thus intimately related to managerial accountability.

The next reform focuses on creating a financial statement that accounts for the capital invested in roads, the impact on this of new investment, and shortfalls in regular road maintenance. It has two parts. The first estimates the total book value of the road network at the end of the fiscal year. This value can be estimated either in great detail, as was done in Hungary (Hungary, Ministry of Transport 1996), or on an approximate basis. An approximate estimate is acceptable if the

results are to be used for illustrative purposes only. It is done by multiplying the length of each type of road by its estimated replacement cost, adding any required inflation adjustment to bring book values to their current replacement costs, and adding any new investment completed during the year. This calculation gives the total book value at the end of the year, set at current replacement costs.

The second part of the statement measures the erosion of capital. It is made up of four items: the rehabilitation backlog at the beginning of the fiscal year (the length of road classified as being in poor condition, multiplied by the average costs of rehabilitating such roads); the amount of rehabilitation completed during the year; the shortfall in regular recurrent maintenance during the year (routine and periodic maintenance); and the additional costs of future road rehabilitation caused by shortfalls in recurrent maintenance (recall chapter 2 suggested that cuts in road maintenance increase the future cash costs of rehabilitation by a factor of two to three). Every four to five years the rough estimate of the rehabilitation backlog should be replaced by a more accurate estimate, based on a road condition survey. The sum of these four items provides an estimate of the current rehabilitation backlog. Finally, the above figures can be used to estimate the current value of the road network and the erosion of capital as a percentage of current book values (table 9.3).

Table 9.3 Prototype road asset statement for a road agency
(millions of dollars)

	December 31, 1996	December 31, 1997
Fixed assets		
Total book value at beginning of year[a]	2,030.00	2,035.70
Adjustment for inflation	0.00	0.00
New works completed during the year[b]	5.70	3.90
Total book value at end of year	2,035.70	2,039.60
Erosion of capital:		
Rehabilitation backlog at beginning of year[c]	−670.00	−714.31
Rehabilitation completed during year	14.95	6.94
Shortfall in recurrent maintenance[d]	−29.63	−26.59
Additional rehabilitation costs[e]	−29.63	−26.59
Rehabilitation backlog at end of year	−714.31	−760.55
Current value of road network	1,321.39	1,279.05
Overall erosion of capital (percent)	35	37

a. Book values are calculated using the following replacement costs per km: paved, $250,000; gravel, $50,000; and earth, $20,000.
b. Investment in new roads and upgrading existing roads.
c. Calculated for all roads in poor condition using the following costs per km for rehabilitation: paved roads, $230,000; gravel, $36,000.
d. Required maintenance expenditures based on the following annual costs per km: paved, $4,000; gravel, $1,000; and earth, $400. Shortfall is the difference between actual maintenance expenditures (from income and expenditure statement) and required maintenance expenditures.
e. A rough estimate based on figures given in chapter 2.

The third reform focuses on the development of a better costing system, usually accomplished by setting up some form of cost accounts. Several countries have already done this, including Finland, New Zealand, Norway, and Sierra Leone, and several developing and transition economies are doing the same, including Botswana, Georgia, Pakistan, and Yemen (box 9.5). Cost accounts show how resources are used, the purpose for which they are used, and how well they serve that purpose. In particular, they show how financial performance varies over time, among different parts of the road agency, and between work done in-house and under contract. Cost accounts provide the basic raw materials needed to operate a maintenance management system effectively. The maintenance management system defines the amount of work required, while the cost accounting system estimates the cost and whether it will be cheaper to do the work in-house or under contract. The system must be kept simple, compatible with existing financial

reporting systems, and capable of being operated given existing staffing and other resource constraints.

Controlling the Quality of Road Works

To ensure that they can deliver quality road services, road agencies must have an effective quality assurance system. The modern view of quality assurance is that it must go well beyond the usual emphasis on controlling the technical quality of road works. Road agencies are becoming customer-oriented organizations, thus quality must be measured in terms of what customers want and whether they are willing to pay for it. This means identifying customers, establishing their needs (surveys of customer satisfaction by road agencies are becoming increasingly common), developing a prioritized program to meet these needs, implementing the projects, and then monitoring how well the projects perform.

Box 9.5 Establishing a commercial cost accounting system for roads in Norway

The Norwegian Public Road Administration is in the process of installing a comprehensive cost accounting system as part of its management information system. About 60 percent of new construction and 25 percent of maintenance is contracted out to private firms. The planning of all maintenance programs and the remaining 75 percent of routine and periodic maintenance are done in-house. In keeping with the government's stated intention of continuously evaluating the productivity of the Administration's in-house civil works operations, a new cost accounting system is being developed and implemented. The aim of this system is to ensure that work is carried out as planned and cost-effectively. The commercial cost accounting system, ECOSYS, should satisfy the government's new directives for all public production units, issued by the Ministry of Finance and the Auditor General's office. ECOSYS has been fully operational since January 1, 1998.

The main goals of ECOSYS are to:
• Provide unit costs for work activity and cost centers to determine productivity and efficiency on a project-by-project and annual basis for each responsibility center.
• Monitor expenditures against the budget.
• Monitor procurement.
The system of coding runs as follows:
• Cost centers and responsibility centers: each cost item is coded according to project and location.

• Work activities: activities are coded according to the Standard Code of Processes, including about 60 processes.
• Personnel: all personnel are coded individually by salary and overhead.
• Plant and equipment: each item of equipment is coded according to unit costs from a separate equipment management system outside of ECOSYS. Unit costs of equipment include capital costs.
• Materials and supplies: all materials and supplies are coded separately.
• Overheads: these include the cost of headquarters personnel and the estimated rent of offices and workshops based on an evaluation of assets.

The operative unit at each county office processes data every second week, based on reports from all personnel, reports for each individual machine, and standard cost reports. The major outputs from the system are monthly and annual reports covering:
• A commercial account separated for the production unit in each county showing income, expenditure, and balance.
• A comparison of budgeted unit costs and actual unit costs.
• A comparison of unit costs between force account production units and private contractors.

Source: Prepared by O. Sylte for this study.

Traditionally, quality was ensured through technical audits (for design work) and by appointing consultants to supervise implementation of works—regularly used for work implemented by contractors and occasionally for work done by in-house staff and equipment. But these arrangements are costly, often create an adversarial relationship between the two parties, and do not place responsibility for quality assurance with the party best able to control it.

Hence came the concept of total quality management. The idea is to place responsibility for quality assurance with the designers and implementers of works. They are best able to control the quality of the products they produce. Total quality management requires that they develop their own quality assurance procedures and have them certified by an independent third party. Consistent application of these procedures is randomly checked by the client (the road agency) and the supervising consultant. On each major project the implementing organizations are required to produce a contract quality plan showing how they intend to control quality and how they propose to monitor their quality management systems.

To assist with this process, the International Standards Organization (ISO) has developed a series of standards for controlling the quality of different types of works—the ISO 9000 family of standards (UNCTAD and WTO 1996). ISO 9001 to 9003 cover quality assurance models for assessment under contractual situations, while other members of the family cover quality assurance guidelines, quality system elements, auditing arrangements, and requirements for measuring systems (box 9.6). ISO 9000 defines the system for total quality management in the following terms:

• *Management.* Senior managers are responsible for quality assurance and are expected to produce a quality policy statement, expressing the overall intentions of the organization relative to product quality. Managers are expected to ensure that the policy is understood, that the roles and responsibilities of all staff operating within the quality management system are clearly defined, and that this policy is implemented throughout the organization. Senior managers are responsible for implementing any required corrective actions.

• *Quality (management) system.* The organization must establish documented procedures for managing quali-

ty. Client requirements must be clearly defined and specifications—or a program or project brief—must be agreed upon, setting down the requirements of the quality management system and how they are to be met. Implementation must be subjected to regular procedural audits.

• *Controlling quality of operations.* Procedures are required for controlling the quality of all operational processes. The procedures must define: arrangements for coordinating among teams and between office and site staff; design methods to be adopted; expected outputs from operations (for example, calculations, drawings, and reports); and how modifications and changes in scope are to be controlled and approved.

• *Document control.* The issue, re-issue, and withdrawal of documents and data must be controlled. The documents include quality procedures, instructions from the client, project briefs, calculations, drawings, conditions of contract, computer output, and certificates.

• *Inspection and testing.* The organization is responsible for the quality of all work carried out by subcontractors. The quality control process should allow for the vetting of subcontractors, contract staff, and specialist consultants. Design procedures must include formal checking and approval of all relevant documents and, before handing over completed work, the organization must ensure that all procedures identified

---

Box 9.6 Terminology associated with total quality management

Quality: The characteristics of a product or service that bear on its ability to satisfy stated or implied customer requirements.

Quality (management) system: The organizational structure, procedures, processes, and resources needed to implement total quality management (defined below).

Quality policy: The overall intentions and direction of an organization with regard to quality, as expressed formally by top management.

Quality assurance: All those planned and systematic actions needed to satisfy management that a particular product will meet given quality standards.

Quality control: The operational techniques and activities used to check that quality requirements have been met.

Total quality management: The organization's approach to managing quality.

in the quality plan have been properly implemented. All equipment that might have a significant impact on the final product (for example, survey equipment), must be properly cared for.

• *Records.* Procedures have to be developed for filing, maintaining, and disposing of records. These should include drawings and details of the works along with audit reports and records of how quality was managed.

• *Audits.* All elements of the quality management system must be audited regularly, following clearly documented procedures. Checking and monitoring of work, including procedural audits, should be carried out by independent personnel.

A number of road agencies have already either implemented ISO 9000 quality assurance systems (FinnRA and Rijkswaterstaat in the Netherlands), are in the process of implementing them (Transit New Zealand), or have decided to implement partial quality assurance systems that do not necessarily meet the ISO 9000 standards (Maryland in the United States). The quality assurance systems cover all of the road agency's internal business processes, provision of consulting and contracting services (including materials suppliers), and partnering programs designed to improve implementation of large civil works projects. The main focus is usually on introducing quality assurance systems for physical works suppliers and, to a lesser extent, professional services. But once a road agency starts requiring quality assurance systems from its suppliers, it must impose the same discipline on itself (see box 9.7).

Many consultants in industrial countries are already accredited, or close to being accredited, under the ISO 9001 standard (the specific model for quality assurance in design, development, production, installation, and servicing). Since evaluating tenders for professional services provides a small competitive advantage to firms that have quality assurance accreditation, many consultants have voluntarily sought certification.

Quality assurance for physical works tends to proceed faster, driven by the interests of major suppliers

---

Box 9.7 Quality assurance for internal business processes: the case of Transit New Zealand

Transit New Zealand proposes to have all of its internal key business processes quality assured by 2000. The key document setting out quality assurance requirements is the Project Management Quality Manual, which deals with the delivery of consistent projects to Transfund, Transit New Zealand's main client. The projects are defined in broad terms related to the delivery and administration of maintenance and capital improvement projects. But project quality plans can be applied to a wide range of projects, including large capital projects, one-off minor projects, projects of a recurring nature, or projects with a more administrative character.

One chapter in the Project Management Quality Manual sets down proformas for project quality plans for simple, standard, and complex projects. These proformas are downloaded from a master file and filled in to suit each project. They provide the basis of the initial project scoping and help to prepare the request for tender.

The development of project quality plans for all projects involving external suppliers became a standard operating procedure in early 1997. The quality assurance audits were initially carried out in selected regional offices by second parties comprising peers from other regional offices. The emphasis was on constructive auditing and process improvement, rather than on trying to catch staff who were not implementing their procedures fully. This constructive style of auditing provided invaluable learning opportunities for both the auditors and the staff being audited. The quality assurance procedures were developed primarily by regional office personnel to ensure that the staff understood the procedures, had an opportunity to suggest improvements, and accepted them as fair and reasonable. An external consultant provided some guidance to ensure that the procedures would meet ISO 9001 requirements.

The second party audits are expected to continue during 1997, becoming progressively more rigorous as the project quality plans and adherence to quality systems becomes more widespread throughout Transit New Zealand's regional offices. A third-party audit by a suitable accreditation agency will initially be undertaken in the first half of 1998 to identify any gaps in the proposed procedures, with the expectation that a full audit for the purposes of accreditation to ISO 9001 will occur late in 1998.

Implementation of the above quality assurance system covering 150 in-house staff cost about $35,000 for the external consultants who helped to develop the quality assurance system, plus most of the time of one staff member for nearly two years. Recurrent costs of operating the system are less, and operation is done primarily by project managers in regional offices as part of their regular jobs.

Source: Prepared by David Rendall for this study.

and of the road agency itself. From the road agency's perspective, quality assurance certification delegates responsibility for achieving quality to the parties best able to manage the associated risks, and it also reduces the need for third-party supervision. Typically, a quality assurance system may increase bid prices by about 2 percent and reduce supervision by about 30 percent (if the supervision fee is 10 percent of the cost of civil works, the saving in supervision costs comes to about 3 percent of the cost of civil works).

The level of quality assurance depends on the project's complexity and size. In New Zealand quality assurance certification operates at two levels. About a quarter of Transit New Zealand's contracts require prior third-party certification of contractors to the higher ISO 9002 standard (the specific model for quality assurance in production, installation, and servicing). The remainder do not require prior certification—the contractor simply develops and works according to a contract quality plan after the contract has been awarded. These quality plans are considered commercially sensitive and are not readily available. In Finland the quality assurance system is designed primarily to produce uniform quality and to encourage the contractors to produce high quality products. It does this by using the individual quality measurements to produce an overall quality index, which is in turn used to compute a quality bonus worth up to 5–6 percent of the contract price. Actual bonuses have been running at about 1–2 percent of the contract price.

Partnering offers a slightly different approach to quality assurance in that it is more concerned with the quality of design and implementation, particularly when projects may have adverse environmental effects and hence low resident support. It is usually done on a voluntary basis, and the road agency tries to ensure that it is not seen to be driving the process for its own ends. In New Zealand partnering is used for major projects and involves an initial one and a half day workshop for all key stakeholders. The stakeholders usually include the client (Transit New Zealand), consultants, contractors and subcontractors, affected property owners, the concerned local authority, and the traffic police. The workshop is managed by a professional facilitator and has to produce a partnering monitoring matrix that documents the shared objectives of the stakeholders,

provides a basis for regular reviews, lists detailed actions required, and includes the partnering charter. Each of the stakeholders typically displays these charters in public places and in their offices. The system often improves project design, encourages local ownership of the project, and may even reduce costs through better collaboration between the consultant and contractor.

In Maryland in the United States partnering was introduced for similar reasons, but with a greater emphasis on winning public support and reducing claims. It is again voluntary, with the road agency inviting the contractor, subcontractors, and suppliers to join in a voluntary partnering agreement. Federal and local government agencies are also invited to join as needed. The arrangement betters communication, speeds up decisionmaking, helps resolve disputes, and expedites implementation. It has also increased innovation, particularly in relation to stakeholder interests. The most spectacular impact of the Maryland partnering program has been on the contractor's claims for payments over and above the original bid price. On nonpartnered projects claims were running at nearly 20 percent of the bid price (settled at slightly more than 5 percent), while on partnered projects they are running at zero. Partnered projects also show impressive savings in terms of the extra work time spent with extra work costs running at less than half those of nonpartnered projects.

Total quality management and partnering are clearly not panaceas that can be applied to all road agencies in all countries. Third-party certification is not an easy thing to achieve—at least using domestic resources—since the skills required are often not available outside of the main road agency itself and can be expensive for smaller-scale contractors. But all road agencies should aspire to a total quality management system and should at least start to consider second-party certification of all internal business processes by their own regional staff and selected third-party certification of major suppliers (such as bitumen suppliers). Another option might be to require third-party certification of contractors as part of the process of registering them as qualified to undertake certain types of road works. Partnering might also be considered in a form adapted to suit local conditions. It offers such important benefits that some attempt should be made to develop a framework applicable at least to very large projects.

Managerial Autonomy and Accountability

The last issue is that of autonomy. Greater autonomy is normally one of the cornerstones of a more commercial approach to management. For more than 20 years the World Bank has been urging governments to grant more autonomy to the managers of parastatals. The objective was to reduce political interference in management decisions, develop a more commercial managerial outlook, reduce overstaffing, and strengthen accountability. The same rationale applies to road agencies. Road managers will not behave commercially until the road agency is more autonomous and managers are held accountable for their performance. And managers cannot be held accountable unless they have sufficient freedom to sign and award contracts, are offered reasonable terms and conditions of employment, and operate without outside interference.

The first step required to strengthen managerial accountability is thus to specify clear objectives and, based on these, set monitorable targets. This should be done in a written document, usually by preparing a corporate plan and using it as the basis for negotiating a performance contract with the parent ministry (box 9.8). In some cases the parent ministry sets targets for the road agency, and the business plan then spells out how the agency intends to meet these targets during the ensuing year (as in the United Kingdom). In others the road agency drafts its own performance agreement and then negotiates the final version with the parent ministry (as in New Zealand). In Ghana the road agency prepares a three-year rolling corporate plan and uses the first year of the plan to draw up a draft performance contract, which is then agreed on with the parent ministry. The performance contracts generally spell out the government's goals for the road agency, strategies for achieving them, and procedures for implementation, monitoring, and control. Monitoring is usually done in terms of the sort of indicators outlined in table 9.2.

Reporting systems are also an important tool for strengthening managerial accountability and should be produced on a regular basis using the sort of indicators included in the above performance contracts. Very few road agencies produce such reports. Most simply produce ad hoc reports when preparing donor-financed

---

> **Box 9.8 Basic principles governing the preparation of contract plans**
>
> The contract plan should be developed jointly by the road agency and the government, and formally ratified by both. It is primarily an implementing document, not a planning document, and will usually be based on the road agency's corporate plan or similar statement of corporate intentions.
>
> It should take the form of a clearly written document ratifying and committing both the road agency and the government to the road agency's objectives and policy choices defined in its corporate plan. It should clarify who has the authority to make decisions, clearly specify those areas where government review or approval is necessary, and set down the road agency's performance goals (in terms of road conditions, staff productivity, and financial targets).
>
> The performance goals should be simple, mutually consistent, and restricted to those items that define the direction of development and measure the performance of senior management. The contract plan should also include a statement of related government commitment, which may include budgetary support, regulatory changes, and potential changes in labor laws and procedures.
>
> *Source:* Prepared by L. Thompson for this study.

---

road projects or during annual and mid-term reviews of such projects. Or, they produce reports that are designed primarily to serve public relations needs.

The annual report should cover such topics as the road agency's mission, its main policies, any changes in legislation, the core road program, maintenance strategies, donor funding, personnel policies, and community participation. It may also include sections on special topics like labor-based road works, development of the local construction industry, and operation of toll roads. The annual budget, the annual statement of accounts, and the auditor's report on the accounts must also be covered. The main road agencies in countries like Finland, Namibia, New Zealand, Romania, South Africa, and the United Kingdom produce fairly comprehensive annual reports, either on their overall operations or on their financial performance. All road agencies should do the same.

Effective auditing is also an important tool for strengthening managerial accountability. Most auditing is done by the government audit office that checks, on a sample basis, information on the amounts and disclosures in the road agency's financial statements. They also assess whether the accounting policies used are appropriate, consistently applied, and adequately disclosed. A few countries are also starting to use private sector auditors to carry out the financial audit, or to use well-known firms of international auditors (as in Latvia, Romania, and Sierra Leone).

These auditors generally do not carry out any technical auditing, although such auditing is often completed as part of the road agency's internal audit function, or is done on an ad hoc basis by the government audit office. More consistent technical and financial auditing, along the same lines as that done for many road funds, would be an improvement. Only then can one be sure that the funds disbursed have been spent on the approved expenditure program and the work has been done according to specifications.

Key Conclusions and Recommendations

The above discussion leads to the following general conclusions and recommendations:

• In terms of both assets and turnover, main road agencies and toll road authorities are well up among the *Fortune* Global 500. The Japan Highway Public Corporation manages assets equal in volume to those of General Motors and has a turnover comparable to Dow Chemicals. Roads are thus big business and deserve to be managed as such.

• Public sector road agencies are likely to function more efficiently when they are faced with some form of competition or a competition surrogate. Strategies to improve managerial performance thus need to concentrate on strengthening market discipline and providing managers with incentives to operate efficiently. The corollary is that managers need to work within an organization that can respond to market discipline.

• The first task is to state clearly the road agency's corporate mission: what is it supposed to be doing and for whom? This is usually set down in a vision statement that expresses a desire to serve the road user, the environment, and the taxpayer by making the road network safer, more reliable, more environmentally acceptable, and more efficient. In the context of this vision statement the parent ministry is increasingly setting a series of quantified performance targets that the road agency is expected to meet.

• Most countries are actively trying to separate planning and management of roads from the implementation of road works to reduce conflicting responsibilities and subject implementation of road works to more market discipline. They have generally either assigned the procurer and producer functions to separate divisions, divided the road agency into separate client and civil works orzanizations, or have contracted out all producer functions.

• When work is contracted out, the road agency has to choose between different forms of contracts and the type of specifications to be used. With regard to the form of contract, there is a choice between lump-sum, admeasure, cost-reimbursable, and target-cost contracts. Most road agencies in developing and transition countries favor price-based contracts, particularly admeasure contracts.

• With regard to specifications, there is a choice between procedural and functional specifications. With procedural specifications, the client defines what work is to be carried out. These specifications are traditionally used for roads, however, they require a lot of supervision and provide few incentives to encourage the contractor to innovate. These have led to a gradual move toward functional specifications, in which the client defines the desired level of service required and leaves the contractor to find the best way of meeting the performance requirements. Functional specifications have resulted in large cost savings, but are probably too complicated for most developing and transition economies.

• Other innovations in contracting procedures include the letting of combined routine and periodic maintenance contracts, multi-year maintenance contracts, and contracts covering several roads within an entire geographical area. Concessions to design, build, finance, and operate selected roads are now being let for periods of 15 to 30 years. Several other changes in contracting procedures are also being tried to encourage innovation. Some contracts require the contractor to design, construct, and guarantee a road pavement for a specified period of time, while on heavily trafficked roads inducements are being offered to encourage the contractor to complete road works as quickly as possible.

• Efforts to contract more work out to the private sector generally must be accompanied by initiatives to develop the local consulting and construction industries and train road agency staff about preparation of bid documents, contracting procedures, contract law, and arbitration procedures.

• The emphasis on establishing a more comrnercially oriented road agency and contracting out means that the road agency will need fewer staff with different qualifications. A fairly efficient road agency should be able to plan and manage the road network with about five or less staff per 100 km. On networks with heavy traffic, the number may rise to 10 per 100 km. The balance of the work would be done by contractors, who might employ about 10 staff per 100 km using capital- intensive construction and maintenance techniques, or 50 per 100 km using labor-intensive techniques.

• Any reduction in the size of the road agency must be accompanied by improved terms and conditions of service, particularly for managers and older staff with experience. Restructuring and downsizing are likely to create redundancies. This is usually dealt with by offering incentives to staff to take early retirement, separating all commercial activities into specialized units and helping the staff to form separate commercial enterprises, assisting in-house laborers to become small-scale contractors, and offering exit packages to the remaining redundant staff.

• Many road agencies are in the process of restructuring their management systems to create a more commercial system with a chief executive and five or six line managers responsible for the main line functions. Staffing structures have been simplified, layering has been reduced, new job descriptions have been prepared to reflect the new responsibilities of each section, and new career and reward systems have been introduced. Most road agencies have also opted for a decentralized regional structure to place staff closer to the customers they serve.

• Road agencies are working actively to develop road management systems to enable them to generate and evaluate alternative ways of maintaining, improving, and extending the road network. The systems should not be too complicated. They should be affordable, suit the decisionmaking needs of the road agency, be compatible with the scarce manpower resources needed to operate them, and be capable of being incrementally upgraded when resources permit. All road agencies should at least have a visual inspection and monitoring system that can be used to raise awareness of maintenance needs.

• Standard government accounting systems, which focus almost exclusively on cash accounting, do not present a clear picture of the road agency's overall financial health and are not capable of producing the financial data needed to plan expenditures and account for the way funds are used. A number of road agencies are therefore restructuring their accounting systems along commercial lines. They are generally moving toward regular commercial accounting systems with income statements, balance sheets, and cash flow statements. Many of the benefits of commercial accounting systems can be achieved with relatively simple reforms that lie within the capacity of most road agencies.

• A relatively simple financial innovation involves preparing a financial statement that clearly shows the amount of capital invested in roads, the impact of new investment, and—most important—shortfalls in regular maintenance. A more extensive reform focuses on developing better costing systems based on some form of cost accounts. Cost accounts show how resources are used, for what purpose, and how well they serve that purpose. Several countries are developing cost accounts to support their new commercial accounting systems.

• To ensure they can deliver quality road services, road agencies need to have an effective quality assurance system. The traditional way of ensuring quality was through technical audits. However, these are costly, often create an adversarial relationship between the parties, and do not place responsibility for quality assurance with the party best able to control it. Hence the concept of total quality management—responsibility for quality assurance rests with the organizations carrying out the works, that is, with the designers and implementers. Total quality management requires that they develop their own quality assurance procedures and have them certified by an independent third party. Application of these procedures is then randomly checked by the client (the road agency) and the supervising consultant.

• The International Standards Organization (ISO) has developed a series of standards for controlling the quality of different types of works. A number of road agencies have already either implemented ISO quality assurance systems, are in the process of implementing them, or have decided to implement partial quality assurance systems that do not necessarily meet the ISO standards. The systems tend to cover quality assurance systems for all the road agency's internal business processes, quality assurance systems covering provision of consulting and contracting services (including materials suppliers), and partnering programs designed to improve implementation of large civil works projects.

• Total quality management is not a panacea that can be applied to quality control in all countries. Third-party certification is not an easy thing to achieve, since the skills required are often not available outside of the main road agency itself and can be expensive for smaller-scale contractors. However, all road agencies should aspire to an eventual total quality management system and might at least start to consider second-party certification of all internal business processes by their own regional staff and selected third-party certification of major suppliers. Another option might be to require third-party certification of contractors as part of the process of registering them as qualified to undertake certain types of road works.

• Partnering offers a slightly different approach to quality assurance in that it is more concerned with the quality of design and implementation, particularly when projects may have adverse environmental impacts and hence low resident support. It is usually done on a voluntary basis, and the road agency tries to ensure that it is not seen to be driving the process for its own ends. The system often improves the project design, encourages local ownership of project, and may even reduce costs through better collaboration between the consultant and contractor.

• Finally, there is the question of managerial autonomy, which is one of the cornerstones of a more commercial approach to management. Managers cannot be held accountable unless they have sufficient freedom to sign and award contracts, are offered reasonable terms and conditions of employment, and operate without outside interference. Many countries are therefore turning their main road agencies into more autonomous arm's-length agencies operating under an annual performance agreement with the parent ministry. To strengthen accountability, road agencies are also being required to publish indicators to measure their performance and to publish comprehensive annual reports. More rigorous technical and financial auditing is also being used.

Notes

1. "The Global 500." *Fortune*, August 5, 1996.
2. The estimated value of the assets of some other road agencies are: Transit New Zealand, $5 billion; the Public Roads Department in Hungary, $4 billion; FinnRA, $25–28 billion; French Directorate of Roads, $132 billion; and Japan Road Bureau, $235 billion.
3. Weak governance plagues many road contracts. Variation orders are a favored device for generating gratification payments and can increase contract costs by as much as five times over the initial bid price.
4. Payment includes an annual lump sum paid in monthly installments, a supervision fee for discrete maintenance, improvement or joint maintenance-improvement activities, and time charge fees for design and preparation of contract documents, maintenance work valued at less than $150,000, and design and preparation of contract documents for all improvement works.
5. These contracts are known as design, build, finance, and operate contracts.
6. During the construction period, the Finnish design, build, finance, and operate contract includes penalties for neglecting to provide a traffic management plan, a fixed sum per day for full closure of the road, a lane closure fee during maintenance periods paid per 10 minute period, and a lane closure fee for every day the lane is closed.
7. The *operation revenue* is all the revenue derived from the annual production contract with FinnRA, plus revenue from work done for outside clients. The *operating result* is the revenue less expenses and depreciation. *Investment* is the capital invested in equipment, buildings, borrow pits, property, stockpiled materials, and work in progress, less current liabilities, which is the sum of all credits for goods received, but not yet paid for. The *gross margin* is the operation revenue less operating expenses.
8. When the Ministry of Works in New Zealand was commercialized and required to prepare regular commercial accounts, it was astonished to learn how much land and other assets it owned and how much these assets were worth. FinnRA has likewise identified that it owns 6,300 pieces of land (1,100 are gravel pits and stone quarries) and 1,800 buildings on another 550 pieces of land. Thirty percent are surplus to requirements, are being sold, and are expected to realize at least $70 million.

Part IV.
Annexes

Annex 1. Length of Road and Estimated Asset Values in Selected Countries

(kilometers)

Country	Total network length	Main paved length	Main unpaved length	Secondary paved length[a]	Secondary unpaved length	Tertiary paved length[b]	Tertiary unpaved length	Total asset value[c] (US$ million)
Argentina	115,810	30,912	5,893	36,389	42,616	0	0	13,426
Chile	35,111	11,559	9,265	692	13,595	0	0	4,144
Ghana	14,750	6,000	8,750	0	0	0	0	1,665
Hungary	29,600	6,800	0	22,800	0	0	0	4,238
Indonesia	69,238	27,326	0	41,912	0	0	0	11,148
Jordan	7,041	2,843	0	2,116	0	2,082	0	820
Kazakhstan	87,523	15,876	1,186	40,063	30,398	0	0	10,391
Korea, Rep. of	31,194	12,052	60	12,508	4,810	0	0	4,870
Pakistan	54,843	6,580	0	48,263	0	0	0	5,739
Russia	465,895	35,978	4,223	238,631	187,063	0	0	41,963
South Africa	331,266	6,100	33	55,564	269,569	0	0	21,017
Thailand	51,126	43,428	7,698	0	0	0	0	10,707
Uruguay	8,629	3,475	0	2,268	1,855	206	825	1,080

Note: Earth roads are not included for any country.

a. Most often these are regional or provincial roads.

b. Only those tertiary networks that are managed by a national road agency.

c. Value of bridges included, taken as additional 5 percent of total value of road network.

Annex 2. The Inverse Elasticity Rule

This Annex presents a simple exposition of the inverse elasticity rule as it might be used to determine an optimal set of road-user charges. The question is how to mobilize a given amount of revenue from each group of road users (cars, buses, trucks) in a way that minimizes overall welfare loss by all user groups. Heuristically, this problem involves minimizing the overall loss of welfare suffered by all road users by equalizing the deadweight loss per dollar of revenue raised from each user group.

The rule will be illustrated in terms of a simple example which assumes that short-run marginal costs of road use are constant (that is, there is no congestion), cross-price elasticities are small enough to be ignored (that is, the travel demand for each user group is independent of the demand of the other user groups), and that relevant elasticities are compensated demand elasticities (see figure A2.1). When the price of road use is raised from P (where it is equal to vehicle operating costs plus the short-run marginal costs of road use) to P', the deadweight loss per dollar of revenue raised, S, is equal to the triangular area ABC divided by the additional net revenue raised, $DCAE$. In other words:

$$S = -\tfrac{1}{2}(\Delta P * \Delta N)/(\Delta P * N') = -\tfrac{1}{2}\Delta N/N'$$
where $\Delta P = (P' - P)$, $\Delta N = (N - N')$.

Since the compensated own-price point elasticity of demand e^A evaluated at point A is defined to be:

$$(\Delta N/N')/(\Delta P/P'),$$

S can be rewritten as:

$$S = 0.5 e^A (\Delta P/P').$$

The overall loss of welfare is minimized by equating S across all user groups:

$$(1) \quad S = e_1{}^A T_1 = e_2{}^A T_2 = \ldots = e_n{}^A T_n,$$

where S represents the welfare gain associated with relaxing the revenue constraint, 1, 2, ... n represent the different user groups, and T_1, T_2, ... T_n represent the relative mark-up of price over the final gross price ($\Delta P_i/P_i'$).

This is the familiar inverse elasticity rule: The ratio of the relative mark-up of user group 1 over user group 2, T_1/T_2, is inversely proportional to the ratio of their respective own-price elasticities of demand, $e_2{}^A/e_1{}^A$. The solution is illustrated in figure A2.2. Note that with a constant demand elasticity, the lines representing group 1, group 2, and group n are straight; otherwise

Figure A2.1 Loss of consumer surplus when price is raised

Price of road use (VOCs, plus user charge)

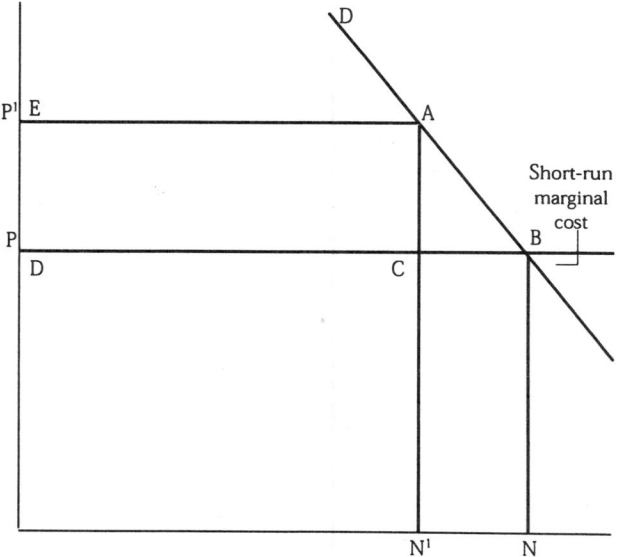

Volume of traffic per time period

Figure A2.2 Equalizing the deadweight loss per dollar of revenue raised

Deadweight loss per dollar of tax revenue, S

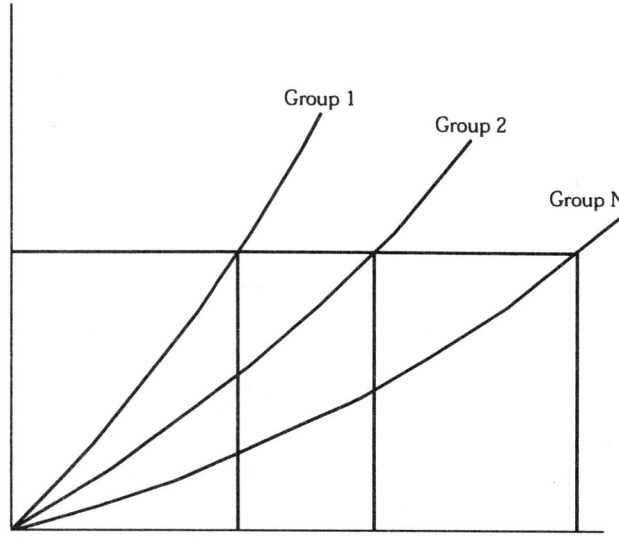

Relative price markup T = Δ P/P¹

they are curved.

The revenue generated by the above mark-ups is:

$$(2) \quad R = T_1 P_1 'N_1' + T_2 P_2 'N_2' + ... T_n P_n 'N_n'$$

$$= \Delta P_1 N_1' + \Delta P_2 N_2' + ... + \Delta P_n N_n',$$

where $N_1' ... N_n'$ represents the volume of each type of traffic at the final traffic levels and $\Delta P_i = P_i * [T/(1 - T_i)]$, $i = 1,...,n$.

Since the values of P, e^A, and R are known, the only unknowns are the values of N' and T. These are estimated from equations 1 and 2 using trial and error or a simple numerical algorithm.[1]

Empirical estimates of the price elasticity of demand for transport generally ignore income effects. When the income effect is thought to be important, the compensated demand elasticity should be used. It is equal to the ordinary demand elasticity plus the proportion of the household budget spent on transport multiplied by the income elasticity of demand for transport. When the cross-price elasticities of demand between the different user groups are significant, the relevant cross-price elasticities should be subtracted: $e^A = (e_{11}{}^A - e_{21}{}^A)$.

In practice, empirical estimates of the price elasticity of demand by different road users are subject to wide margins of error (see Oum, Waters, and Yong 1990). Recent estimates vary from: 0.10 to 1.1.0 for a car, 0.10 to 1.30 for a bus, to 0.70 to 1.10 for a truck. This variance reflects the fact that demand elasticities depend on market conditions, which vary widely throughout the road network. It is therefore unwise to use average or typical demand elasticities to estimate road-user charges. Instead, it is better to use uniform mark-ups (that is, to assume demand elasticities are equal) and to use differential mark-ups only when accurate and consistent country-specific values are available.

Finally, when roads are congested and short-run marginal costs are not constant, the analysis must include the supply elasticities, which greatly complicates the analysis.

Note

1. The trial and error method goes as follows: Choose a starting value for S and solve equation 1 for T_1. Estimate $\Delta P_i = P_i * [T/(1 - T_i)]$. Assume N_i' is approximately equal to N_i. Calculate the implied value of R from equation 2 and compare it to the actual value of R. If the implied value is less than the actual value, choose a higher value for S and repeat the calculation. The values converge after three to five iterations. Finally, check whether ΔP_i is large enough to make N_i' significantly lower than N_i. If so, replace N_i with a new estimate of N_i' and repeat the above calculations.

Annex 3. Estimating Road-User Charges: A Worked Example

This annex takes a hypothetical road network and, using the pricing and cost recovery policies developed in chapter 7, estimates the user charges required to ensure that: the costs of operating and maintaining the main trunk roads managed by the main road agency are fully funded, the grants made to road agencies managing urban and rural roads are sufficient to ensure that their maintenance programs are also fully funded, and sufficient funds are available to finance investment in main roads, support investment in urban and rural roads, and meet debt service obligations. The estimates were prepared using the Road User Charges Model, version 2.0, developed by Ian Heggie and Rodrigo Archando-Callao.[1]

The hypothetical road network consists of 8,550 km of trunk roads managed by the main road agency (7,500 km are paved); 66,000 km of rural roads managed by the main road agency, states, provinces, or rural district councils; and 5,300 km of urban roads managed by urban district councils or municipalities (see table A3.1). Traffic volumes vary from 300 vpd to 10,000 vpd on paved roads (with 30 percent trucks), from 50 vpd to 300 vpd on gravel roads, and are 25 vpd on earth roads. The vehicle fleet consists of 351,000 vehicles (see table A3.2). About 8 percent are trucks and buses, which account for more than 20 percent of annual vehicle-kilometers. Average annual distances traveled vary from 18,000 to 25,000 km for cars, to 50,000 km for most trucks, to 80,000 km for taxis, buses, and articulated trucks.

There are six main steps to the analysis:
- Review the road agency's annual financial needs for annual (routine and recurrent) and periodic maintenance set out in attachment 1. It contains default road class descriptions and unit maintenance needs for each road class computed using the World Bank's HDM model, which is based on data collected from developing and transition economies worldwide. The figures in the attachment represent medium-size VOCs, agency costs, and climatic (environment) conditions. If necessary, adjust or change the default unit annual maintenance needs required to maintain each road class on a sustainable basis.[2]
- Estimate the costs of operating and maintaining the entire road network on a sustainable basis. The entire road network includes main trunk roads managed by the main road agency; rural roads managed by the main road agency, states, provinces or rural district councils; and urban streets and avenues managed by urban district councils or municipalities (see table A3.1).
- Using the above costs, prepare an outline financing table (table A3.3) defining all yearly needs for maintenance and investment to establish which costs have to be met through user charges. These costs will include the entire costs of the main trunk road network and part of the costs associated with rural and urban roads. The remaining costs are assumed to be met with local revenues (such as parking charges and local property taxes).
- Define the characteristics of vehicles using the road network (table A3.2) and compute the variable costs attributable to each type of vehicle (table A3.4). Table A3.2 also provides an estimate of total annual fuel consumption.
- Enter the current fuel levy, along with the annual license fees and, where relevant, the axle-loading charge (that is, supplementary heavy vehicle license fees) to ensure that: each vehicle class covers its variable costs and all vehicles together generate sufficient revenues to cover all the costs included in the outline

Table A3.1 Yearly agency needs for maintaining the road network on a sustainable basis

Road network	Road type	Traffic (vpd)	Description	Length (km)	Vehicle utilization (M veh-km/yr)	Annual maintenance (M$/yr) Fixed	Variable	Total	Periodic maintenance (M$/yr) Fixed	Variable	Total
Main roads:	*Paved*	300	30% Trucks/Loading: Low	3,400	372	3.40	0.08	3.48	7.21	4.12	11.33
Trunk roads		600	30% Trucks/Loading: Low	1,890	414	1.89	0.09	1.98	6.45	2.37	8.82
managed		1,000	30% Trucks/Loading: Low	880	321	0.88	0.05	0.93	3.77	0.74	4.51
by main road		3,000	30% Trucks/Loading: Low	1,150	1,259	1.15	0.14	1.29	6.85	1.20	8.05
agency		6,000	30% Trucks/Loading: Low	110	241	0.11	0.02	0.13	0.68	0.12	0.80
		10,000	30% Trucks/Loading: Low	70	256	0.07	0.01	0.08	0.44	0.10	0.53
			Total	7,500	2,863	7.50	0.40	7.90	25.41	8.64	34.05
	Gravel	50	Importance: Primary	450	8	0.23	0.14	0.36	0.41	0.04	0.45
		100	Importance: Primary	380	14	0.19	0.17	0.36	0.35	0.08	0.42
		200	Importance: Primary	180	13	0.09	0.11	0.20	0.16	0.09	0.26
		300	Importance: Primary	40	4	0.02	0.03	0.05	0.04	0.03	0.07
			Total	1,050	40	0.53	0.44	0.97	0.95	0.24	1.20
	Total			8,550	2,903	8.03	0.84	8.87	26.36	8.88	35.24
Rural roads:	*Paved*	300	30% Trucks/Loading: Low	2,000	219	2.00	0.05	2.05	4.24	2.42	6.67
Rural roads			Total	2,000	219	2.00	0.05	2.05	4.24	2.42	6.67
managed by	*Gravel*	50	Importance: Primary	10,000	183	5.00	3.00	8.00	9.09	0.91	10.00
main road		100	Importance: Primary	10,000	365	5.00	4.50	9.50	9.09	2.02	11.11
agency, states,		200	Importance: Primary	4,000	292	2.00	2.40	4.40	3.64	2.08	5.71
or district			Total	24,000	840	12.00	9.90	21.90	21.82	5.01	26.83
councils	*Earth*	25	Importance: Primary	40,000	365	10.00	6.48	16.48	n.a.	n.a.	n.a.
			Total	40,000	365	10.00	6.5	16.5	n.a.	n.a.	n.a.
	Total			66,000	1,424	24.00	16.43	40.43	26.06	7.43	33.49
Urban	*Paved*	300	30% Trucks/Loading: Low	1,700	186	1.70	0.04	1.74	3.61	2.06	5.67
streets and		600	30% Trucks/Loading: Low	300	66	0.30	0.01	0.31	1.02	0.38	1.40
avenues:		1,000	30% Trucks/Loading: Low	300	110	0.30	0.02	0.32	1.29	0.25	1.54
Urban roads		3,000	30% Trucks/Loading: Low	1,200	1,314	1.20	0.14	1.34	7.15	1.25	8.40
managed by		6,000	30% Trucks/Loading: Low	500	1,095	0.50	0.09	0.59	3.11	0.53	3.65
district		10,000	30% Trucks/Loading: Low	400	1,460	0.40	0.08	0.48	2.49	0.55	3.04
councils or			Total	4,400	4,230	4.40	0.39	4.79	18.67	5.03	23.69
municipal-	*Gravel*	50	Importance: Primary	900	16	0.45	0.27	0.72	0.82	0.08	0.90
ities			Total	900	16	0.45	0.27	0.72	0.82	0.08	0.90
	Total			5,300	4,247	4.85	0.66	5.51	19.48	5.11	24.59

n.a. Not applicable.

financing table (table A3.3). These calculations are shown in tables A3.5–A3.7.

• Find an optimal set of user charges (the axle-loading charge and the fuel levies), using the Solve option in Excel, that equilibrates revenues and financial needs, while also minimizing the surplus of road-user revenues over costs. These calculations are shown in tables A3.8–A3.10.

Finally, tables A3.11 to A3.13 provide a summary of yearly road agency needs, how they might be financed, the road-user revenues generated by current user charges and the optimal charges, and the current unit road-user charges and the optimal charges. Table A3.11 shows that total road agency needs come to $227.78 million, including $54.8 million for annual maintenance (24 percent), $93.33 million for periodic maintenance (41 percent), and the balance of $79.65 million for investment (35 percent). Of this total, $141.14 million would be financed through road-user charges (62 percent) and the balance through local revenues. The optimal road-user charges required to generate these revenues vary from 1.47 cents per km for a car (gasoline), to 1.14 cents per km for a light truck and 1.86 cents per km for a bus, to 4.21 cents per km for an articulated truck.

The optimum axle-loading charge works out to be $96.47 per ESA per year, while the fuel levies work out

Table A3.2 Characteristics of vehicles using the road network

Vehicle type	Number of vehicles	Kilometers driven per year (km/yr)	Equivalent standard axle per vehicle (ESA/veh)	Fuel consumption (l/ veh-km)	Vehicle utilization veh-km/yr (million)	Loading impact ESA-km/yr (million)	Fuel consumption l/yr (million)
Car, gasoline	220,000	18,000	0.000	0.08	3,960	0	317
Car, diesel	80,000	25,000	0.000	0.10	2,000	0	200
Taxi, gasoline	0	0	0.000	0.08	0	0	0
Taxi, diesel	4,000	80,000	0.000	0.10	320	0	32
Utility	20,000	35,000	0.001	0.09	700	1	63
Light truck	2,000	50,000	0.100	0.15	100	10	15
Medium truck	11,600	50,000	1.250	0.20	580	725	116
Heavy truck	3,600	70,000	3.000	0.25	252	756	63
Articulated truck	6,000	80,000	5.000	0.35	480	2,400	168
Bus	4,400	80,000	0.500	0.25	352	176	88
Total[a]	351,600				8,744	4,068	1,062

a. Vehicle utilization from table A3.1 is 8,573 million veh-km/yr, which corresponds closely with the figure of 8,744 million veh-km/yr produced by this table.

Table A3.3 Yearly needs and financing subdivided into main cost components

(million dollars per year)

Expenditure type			Yearly needs Recurrent Fixed costs	Yearly needs Recurrent Variable costs	Investments	Total
Recurrent expenditures	Main roads	Annual maintenance	8.03	0.84		8.87
		Periodic maintenance	26.36	8.88		35.24
		Traffic enforcement[a]	2.35	1.00		3.35
		Administration[b]	4.40	1.90		6.30
		Interest charges[c]	5.50	0.00		5.50
		Total	46.64	12.62		59.26
	Rural roads	Grants for maintenance[d]				
		Annual maintenance	24.00	16.43		40.43
		Periodic maintenance	26.06	7.43		33.49
		Total	50.06	23.86		73.92
	Urban streets and avenues	Grants for maintenance[d]				
		Annual maintenance	4.85	0.66		5.51
		Periodic maintenance	19.48	5.11		24.59
		Total	24.33	5.77		30.10
	Total		121.03	42.25		163.28
Investments	Main roads				25.00	25.00
	Debt service / repayment				15.00	15.00
	Grants for rural roads[e]				4.50	4.50
	Grants for urban streets and avenues[e]				20.00	20.00
	Total				64.50	64.50
Total			121.03	42.25	64.50	227.78

Variable costs that vary with vehicle utilization (annual maintenance, traffic enforcement, administration and interest)   20.83

Variable costs that vary with axle loading (periodic maintenance)   21.42

   42.25

a. In this example an estimated 70 percent of traffic enforcement costs are fixed.

to be $0.08 per liter for gasoline and $0.05 per liter for diesel fuel. The combined annual vehicle license fee and supplementary heavy vehicle fee work out to be $150 for a car, $621 for a medium-weight truck, $1,289 for a heavy truck, $1,982 for an articulated truck, and $500 for a bus.

			Financing		
	Financed by user charges			*Financed by local revenues*	
Recurrent					
Fixed costs	*Variable costs*	*Investments*	*Total*	*(%)*	*Total*
8.03	0.84		8.87		
26.36	8.88		35.24		
2.35	1.00		3.35		
4.40	1.90		6.30		
5.50	0.00		5.50		
46.64	12.62		59.26	Percentage of fixed	
0.00	16.43		16.43	100	24.00
0.00	7.43		7.43	100	26.06
0.00	23.86		23.86	Percentage of fixed	50.06
0.00	0.66		0.66	100	4.85
0.00	5.11		5.11	100	19.48
0.00	5.77		5.77		24.33
46.64	42.25		88.89		74.39
		25.00	25.00	Percentage	
		15.00	15.00	of total	
		2.25	2.25	50	2.25
				Percentage of total	
		10.00	10.00	50	10.00
		52.25	52.25		12.25
46.64	42.25	52.25	141.14		86.64

b. In this example fixed costs include expenditures on buildings and 70 percent of headquarters salaries.

c. In this example interest charges on road loans are fixed costs.

d. User charges cover all variable costs and, in this example, local revenues cover 100 percent of fixed costs.

e. In this example local revenues cover 50 percent of the grants for investments for rural roads and urban streets and avenues.

Table A3.4 Road-user charges required to cover variable costs

| Vehicle type | Variable cost requirements(M$/yr) | | | Vehicle utilization veh-km/yr (million) | Charges to cover variable cost (c/veh-km) | | |
	Vehicle related	Loading related	Total		Vehicle Related	Loading related	Total
Car, gasoline	9.43	0.00	9.43	3,960	0.24	0.00	0.24
Car, diesel	4.76	0.00	4.77	2,000	0.24	0.00	0.24
Taxi, gasoline	0.00	0.00	0.00	0	0.00	0.00	0.00
Taxi, diesel	0.76	0.00	0.76	320	0.24	0.00	0.24
Utility	1.67	0.00	1.67	700	0.24	0.00	0.24
Light truck	0.24	0.05	0.29	100	0.24	0.05	0.29
Medium truck	1.38	3.82	5.20	580	0.24	0.66	0.90
Heavy truck	0.60	3.98	4.58	252	0.24	1.58	1.82
Articulated truck	1.14	12.64	13.78	480	0.24	2.63	2.87
Bus	0.84	0.93	1.77	352	0.24	0.26	0.50
Total	20.83	21.42	42.25	n.a.	n.a.	n.a.	n.a.

n.a. Not available.

Table A3.5 Current license fees, axle loading charges, and fuel levies

Vehicle type	Standard license fee ($/veh-yr)	Equivalent standard axle (ESA/veh)	Axle loading license fee ($/veh-yr)
Car, gasoline	150	n.a.	n.a.
Car, diesel	200	n.a.	n.a.
Taxi, gasoline	0	n.a.	n.a.
Taxi, diesel	150	n.a.	n.a.
Utility	150	n.a.	n.a.
Light truck	200	n.a.	n.a.
Medium truck	500	1.25	125
Heavy truck	1,000	3.00	300
Articulated truck	1,500	5.00	500
Bus	500		n.a.

n.a. Not available.

Axle loading charg ($/ESA/yr)	100.00
Gasoline levy—gasoline cars and taxis ($/l)	0.10
Diesel levy—diesel cars, taxis, buses, and trucks ($/l)	0.08

Table A3.6 Current unit and total road-user revenues

| Vehicle type | Unit road-user revenues(c/veh-km) | | | | Vehicle utilization veh-km/yr (million) | Total road-user revenues (M$/yr) | | | |
	Fuel levy	Standard license fee	Axle loading license fee	Total		Fuel levy	Standard license fee	Axle loading license fee	Total
Car, gasoline	0.80	0.83	0.00	1.63	3,960	31.68	33.00	0.00	64.68
Car, diesel	0.80	0.80	0.00	1.60	2,000	16.00	16.00	0.00	32.00
Taxi, gasoline	0.80	0.00	0.00	0.80	0	0.00	0.00	0.00	0.00
Taxi, diesel	0.80	0.19	0.00	0.99	320	2.56	0.60	0.00	3.16
Utility	0.72	0.43	0.00	1.15	700	5.04	3.00	0.00	8.04
Light truck	1.20	0.40	0.00	1.60	100	1.20	0.40	0.00	1.60
Medium truck	1.60	1.00	0.25	2.85	580	9.28	5.80	1.45	16.53
Heavy truck	2.00	1.43	0.43	3.86	252	5.04	3.60	1.08	9.72
Articulated truck	2.80	1.88	0.63	5.30	480	13.44	9.00	3.00	25.44
Bus	2.00	0.63	0.00	2.63	352	7.04	2.20	0.00	9.24
Total	n.a.	n.a.	n.a.	n.a.	n.a.	91.28	73.60	5.53	170.41

n.a. Not available.

Table A3.7 Variable costs and total costs

Vehicle type	Charges needed to cover variable costs (c/veh-km)	Current unit road-user revenues (c/veh-km)	Current user revenues > charges needed	Current user revenues charges needed- (c/veh-km)	Variable user cost surplus (M$/yr)
Car, gasoline	0.24	1.63	Yes	1.40	55.2
Car, diesel	0.24	1.60	Yes	1.36	27.2
Taxi, gasoline	0.00	0.80	Yes	0.80	0.0
Taxi, diesel	0.24	0.99	Yes	0.75	2.4
Utility	0.24	1.15	Yes	0.91	6.4
Light truck	0.29	1.60	Yes	1.31	1.3
Medium truck	0.90	2.85	Yes	1.95	11.3
Heavy truck	1.82	3.86	Yes	2.04	5.1
Articulated truck	2.87	5.30	Yes	2.43	11.7
Bus	0.50	2.63	Yes	2.12	7.5
Total	n.a.	n.a.	n.a.	15.07	128.2

n.a. Not available.

Total financing needs (M$/yr)	141.12
Total revenues (M$/yr)	170.41
Revenues - financing needs (M$/yr)	29.27

Table A3.8 Optimal axle loading charge and fuel levy

Vehicle type	Standard license fee ($/veh-yr)	Equivalent standard axle (ESA/veh)	Axle loading license fee ($/veh-yr)
Car, gasoline	150	n.a.	n.a.
Car, diesel	200	n.a.	n.a.
Taxi, gasoline	0	n.a.	n.a.
Taxi, diesel	150	n.a.	n.a.
Utility	150	n.a.	n.a.
Light truck	200	n.a.	n.a.
Medium truck	500	1.25	121
Heavy truck	1,000	3.00	289
Articulated truck	1,500	5.00	482
Bus	500	n.a.	n.a.

n.a. Not available.

Axle loading charg ($/ESA/yr)	96.47
Gasoline levy—gasoline cars and taxis ($/l)	0.08
Diesel levy—diesel cars, taxis, buses, and trucks ($/l)	0.05

Table A3.9 Optimized unit road-user revenues and total road-user revenues

Vehicle type	Unit road-user revenues (c/veh-km)				Vehicle utilization veh-km/yr (million)	Total road-user revenues (M$/yr)			
	Fuel levy	Standard license fee	Axle loading license fee	Total		Fuel levy	Standard license fee	Axle loading license fee	Total
Car, gasoline	0.64	0.83	0.00	1.47	3,960	25.27	33.00	0.00	58.27
Car, diesel	0.50	0.80	0.00	1.30	2,000	9.91	16.00	0.00	25.91
Taxi, gasoline	0.64	0.00	0.00	0.64	0	0.00	0.00	0.00	0.00
Taxi, diesel	0.50	0.19	0.00	0.68	320	1.59	0.60	0.00	2.19
Utility	0.45	0.43	0.00	0.87	700	3.12	3.00	0.00	6.12
Light truck	0.74	0.40	0.00	1.14	100	0.74	0.40	0.00	1.14
Medium truck	0.99	1.00	0.24	2.23	580	5.75	5.80	1.40	12.95
Heavy truck	1.24	1.43	0.41	3.08	252	3.12	3.60	1.04	7.76
Articulated truck	1.73	1.88	0.60	4.21	480	8.33	9.00	2.89	20.22
Bus	1.24	0.63	0.00	1.86	352	4.36	2.20	0.00	6.56
Total	n.a.	n.a.	n.a.	n.a.	n.a.	62.20	73.60	5.33	141.14

n.a. Not available.

Table A3.10 Variable costs and total costs

Vehicle type	Charges needed to cover variable costs (c/veh-km)	Optimized unit road-user revenues (c/veh-km)	Optimized user revenues > charges needed	Optimized user revenues - charges needed (c/veh-km)	Variable user cost surplus (M$/yr)
Car, gasoline	0.24	1.47	Yes	1.23	48.8
Car, diesel	0.24	1.30	Yes	1.06	21.1
Taxi, gasoline	0.00	0.64	Yes	0.64	0.0
Taxi, diesel	0.24	0.68	Yes	0.44	1.4
Utility	0.24	0.87	Yes	0.64	4.5
Light truck	0.29	1.14	Yes	0.85	0.9
Medium truck	0.90	2.23	Yes	1.34	7.7
Heavy truck	1.82	3.08	Yes	1.26	3.2
Articulated truck	2.87	4.21	Yes	1.34	6.4
Bus	0.50	1.86	Yes	1.36	4.8
Total	n.a.	n.a.	n.a.	10.17	98.9

n.a. Not available.

Total financing needs (M$/yr)	141.14
Total revenues (M$/yr)	141.14
Revenues - financing needs (M$/yr)	0.00

Notes

1. The model can be accessed and down-loaded from the World Bank's transport web page located at http://www-int.world-bank.org/fpsi/infra/transport/. It appears in the Rural Roads & Highways knowledge base under Databases & Software.

2. More accurate costs can be computed by using the World Bank's HDM model, or the default values supplied in attachment 1 can be replaced by other default values—reflecting different VOCs, agency costs, traffic composition, loading, and environmental conditions—by using the Excel model (PNEEDS10XLS for paved roads and UNEEDS.XLS for unpaved roads) supplied with the Road User Charges Model.

Annex 4. Review of Selected Road Funds

Japan: Road Improvement Special Account

Japan introduced a special funding system for roads in 1954, coninciding with the introduction of the first five-year road improvement program. At the end of the war there were about 130,000 motor vehicles in Japan, but this figure jumped to one million by 1953, and it became clear that the road network—which was outdated and in poor condition—had to be improved. These five-year rolling programs were designed to bring the Japanese road system into the twentieth century and to adjust to the rapid growth in motor vehicles. Since then, the five-year road improvement programs have been renewed and implemented continuously to provide road users with better driving conditions and to provide people living in urban areas with better access to the countryside.

The new funding system for roads involved earmarking certain road-related taxes and depositing them into a special account, or road fund. This funding system was introduced to meet the needs of the postwar road improvement program and was "based on the concept that road users who enjoy the benefits of improved roads should bear the burden for their improvement" (that is, it was based on the user pay concept).

The road fund employs an elaborate system of earmarked national and local taxes to finance the maintenance, improvement, and construction of national, prefectural, and local roads. At the national level tax revenues earmarked for roads are allocated among the various road authorities as follows:
- Twenty-five percent of the gasoline tax ($0.12 per liter) is transferred to the road improvement special account.
- Half the motor vehicle liquid petroleum gas tax ($0.14 per kg) is paid into the special account, while

the remainder is transferred to local governments as motor vehicle liquid petroleum gas transfer tax.
- Seventy-five percent of the motor vehicle tonnage tax ($51 per half ton per year) is paid into the special account, while the remainder is transferred to local governments as the motor vehicle tonnage transfer tax.

At the local level tax revenues earmarked for roads are allocated among the various road authorities as follows:
- The liquid petroleum gas tax is spent on roads in the Tokyo Metropolitan Area, Hokkaido, prefectures, and designated cities.
- The motor vehicle tonnage tax is spent on roads in cities, towns, and villages.
- Forty-three percent of the local gasoline tax ($0.05 per liter) is spent on roads in the Tokyo Metropolitan Area, Hokkaido, prefectures, and designated cities, while the other 57 percent is spent on roads in cities, towns, and villages.
- The local diesel fuel tax ($0.31 per liter) is spent on roads in the Tokyo Metropolitan Area, Hokkaido, prefectures, and designated cities.
- Thirty percent of the motor vehicle purchase tax (5 percent of the purchase price for private motor vehicles) is spent on roads in the Tokyo Metropolitan Area, Hokkaido, and prefectures, while the other 70 percent is spent on roads in cities, towns, and villages.

Earmarked revenues at both the national and local levels are supplemented by general tax revenues and, in the case of the national government, are also deposited into the Road Improvement Special Account to ensure comprehensive management of the funds. Revenue from user fees in 1995 was roughly $30 billion.

Funds from the Road Improvement Special Account are provided to road authorities on a cost-share basis.

The central government finances half the costs of maintaining directly managed national highways. The remaining costs are financed by prefectural governments and designated large cities. The central government also finances two-thirds of the costs of improving directly managed national highways, 70 percent of the national expressway network, and 50 percent of subsidized national highways, main local (prefectural) roads, and main local (municipal) roads.

Road spending in Japan is based on five-year road improvement programs prepared by the Ministry of Construction. The process worked well up until the start of the Ninth Road Improvement Program. Programs were prepared and approved, and corresponding tax rates were then written into a new proper tax law, which ensured that the road fund generated sufficient funds to cover costs during the next five-year period. But in 1982 a concerted effort was made to abolish the road fund and replace it with allocations from the government's consolidated budget. Although a roads board was in place— the Japan Road Council—up to that point it had played a relatively nominal role relative to the road fund. The role and duties of the Council are laid down in article 77 of the Road Law. The law held that a Council must be established by the Ministry of Construction at the request of the minister. Among other things, the Council is asked to, "deliberate on management of the road fund and on toll road financing and advise the Minister on changes necessary to reorient road financing."

Faced with this crisis of the road fund, the Ministry of Construction asked the Road Council to conduct an inquiry and make recommendations regarding how the overall road network should be developed as the country approached the twenty-first century. Their report, *Proposal for Road Improvement Approaching the 21st Century*, not only set the future direction of the road program, but also saved the road fund and established the credibility of the Road Council. Since then, the Ministry of Construction has always asked the Road Council to submit its views on a long-term strategy for road improvement as part of the preparations for the Five-Year Road Improvement Program.

The Council was established in 1952 and consists of a chairperson and 12 other members. The members are nominated by the director general of roads and are appointed by the minister of construction. The chair-

person has traditionally been the president of Japan Road Association (always a former undersecretary from the Ministry of Construction), but is currently the former president and chairperson of Nissan Corporation. Board members include representatives of the motor industry, business community, trade unions, academia, and local government. Much of the Council's substantive work is carried out by three subcommittees: one deals with road policy, one with toll roads, and the other with environmental issues. The Council has no permanent secretariat, but is serviced by staff from the Roads Bureau of the Ministry of Construction.

Day-to-day management of the road fund is carried out by the General Affairs Division of the Roads Bureau. They have about 12 staff who are responsible for forecasting revenues, liasing with Ministry of Finance, and monitoring use of funds by the other divisions of the Roads Bureau and the prefectures. Each of these divisions (for example, the Expressway Corporation and the Highway Division) and the prefectures have two or more accountants who monitor the expenditure programs and report back to the General Affairs Division. Expenditures on roads in cities, towns, and villages are monitored by the prefectures who then report back to the General Affairs Division on programs supported by the road fund.

The road fund acts like a line of credit. Once parliament has approved the overall spending limits, the Ministry of Construction can draw down the funds regardless of the actual revenue in the road fund account at the central bank (that is, the government provides working capital). Contractors are paid directly after work has been inspected by an experienced Ministry of Construction engineer who has not been involved in planning or implementing the work. Work carried out by prefectures and designated cities is also inspected by Ministry of Construction engineers.

All work financed from the road fund is subjected to an audit by the Japanese Institute of Audits, which is independent of the government and influential amongst the public. The audit is conducted on a sample basis, targeting several specific works per office. The audit team visits the work office, examines control procedures and financial records, and dispatches civil engineers to inspect the selected work sites. Problems and queries are resolved with the Ministry of

Construction and the audit report is then submitted to parliament.

Transfund New Zealand

The original road fund in New Zealand was established in 1953. In 1989 the road fund was renamed the Land Transport Fund and its management fund was transferred to Transit New Zealand, which had been set up in 1989. But since the road fund was used to finance Transit New Zealand's road program, as well as those of the Regional Councils and District Councils, there was thought to be a conflict of interest. Thus on July 1, 1996 the Transit New Zealand Amendment Act came into effect, creating a new agency called Transfund New Zealand (Transfund). Management of the road fund was therefore separated from Transit New Zealand and placed under the jurisdiction of a separate management board.

The board consists of five members: two representing Transit New Zealand (either employees or members of the Transit New Zealand Authority), one representing local government, one representing road users, and one representing other aspects of the public interest. Members are appointed by the governor-general on the recommendation of the responsible minister. The chairperson is appointed by the governor-general from among the existing members of the board.

The revenue for the road fund comes from: a fuel excise added to the price of gasoline, weight-distance charges paid by diesel vehicles, motor vehicle registration fees, interest earned on the road fund account, revenues earned from sale of surplus property, and refund of the GST (the New Zealand equivalent of a value-added tax).

• The fuel excise in 1996 was set at about $0.065 per liter (the total excise tax on gasoline was $0.21 per liter) and is expected to generate about $204 million. The funds are collected by the New Zealand Customs, which is paid about $414,000 (about 0.2 percent of the revenue) to cover their costs. Evasion is negligible since the funds are collected at source at New Zealand's only refinery or at ports of entry.

• Weight-distance charges are expected to generate about $293 million in 1996. The collection is managed by a unit within the Land Transport Safety Authority at a cost of about $10 million (including $5 million spent on enforcement). A number of agencies sell the weight-distance certificates, including the New Zealand Post (approximately 50 percent), BP petrol stations, Vehicle Testing New Zealand, Vehicle Inspection New Zealand, New Zealand Automobile Association, and AMI Insurance. There is also an arrangement whereby operators can buy licenses from their own offices by way of a remote terminal. The overall costs are about $1 to $2 per transaction. Evasion accounts for about 12 percent of revenues (9.4 percent from heavy vehicles and 2.8 percent from light vehicles) and legal avoidance for about 7 percent of net revenues.

• Motor vehicle registration fees are expected to generate about $104 million in 1996. The collection is managed by the Land Transport Safety Authority at a cost of about $19.3 million (nearly 19 percent of total revenues). Similar agencies sell the registration certificates: New Zealand Post, Vehicle Testing New Zealand, Vehicle Inspection New Zealand, the Automobile Association, and AMI Insurance. The extent of evasion is unknown.

• Interest and sale of surplus property are minor items. But payment of interest recognizes that the funds held by the Treasury belong to Transfund and that short-term borrowing must be paid for.

• Reimbursement of GST is at a rate of one-ninth of the expenditures made by the Land Transport Fund to compensate for payment of GST on all revenues received by the road fund.

The main objective of the board is to "allocate resources to achieve a safe and efficient roading system." In this connection its key functions are to:

• Approve and purchase a national roading program from the various road agencies, including capital projects.

• Approve the competitive pricing procedures applicable to the roading program.

• Audit the performance of Transit New Zealand and local authorities against their respective roading programs.

• Provide advice and assistance to local authorities in relation to the new Transfund Act.

Transfund has 36 staff members, including a chief executive who is appointed by the board. The chief executive appoints all other staff members. They include four policy staff, four administrative staff, eight programming and contracts staff, nine audit staff, and ten staff in seven regional offices.

Transfund manages the National Roads Fund, which has been reconstituted from the old Land Transport Fund. The key changes are the new management structure and the removal of the need for separate decisions on the funding level and the expenditure program. The government still sets the charges that determine the inflows to the road fund, but no longer determines the outflows. Once the costs of police and the Land Transport Safety Authority have been met, the balance of the revenues are available for use by Transfund without any further controls. In other words, the charges are still being collected as if they were taxes, but Transfund is now wholly responsible for what happens to the revenues.

The specific responsibilities of Transfund are to:
• Prepare the Annual National Roading Program.
• Recommend to the government income and expenditure levels needed to support the Program.
• Advise on the suitability of the Land Transport system.
• Fund the approved projects within the Program.
• Make payments to road agencies to finance the approved projects.

The National Land Transport Plan is thus the basic building block for Transfund's short-term and long-term activities. It is built up from bids submitted by Transit New Zealand and the local authorities. The bids are subject to checks on the reasonableness and appropriateness of supporting benefit-cost calculations, before projects are ranked in order of priority. Maintenance is accorded highest priority, with other projects ranked in order until all available funds are used (the current cut-off benefit-cost ratio is 4).

Maintenance requirements are based on a combination of professional judgment and the outcome of the Road Assessment Maintenance Management System (RAMM). RAMM is a computerized pavement management system that includes road inventory (road condition) and treatment selection for determining work programs based on engineering and economic criteria. Transfund requires that all road agencies wishing to receive funds from the road fund base their estimated funding requirements on RAMM. Road authority requests are vetted on an ongoing basis by

Transfund staff, and the Review and Audit Division carries out audits every three years to ensure that minimum maintenance standards and service levels are being maintained by each road authority. To further refine the method of allocating maintenance funds, a project has recently been launched to determine the best way of estimating optimal maintenance funding levels for the different road authorities.

The Review and Audit Division carries out systematic reviews and appraisals of activities wholly or partly funded from the road fund. The chief executive reports to the board and, in exceptional circumstances, may report directly to the chairperson. One of the conditions for providing funds to the road authorities is that they provide all the information and cooperation necessary to enable the division to review and audit the correct application of these funds. The aim of the audits is to ensure that the funds have been used in an efficient and effective manner. The division monitors outputs in relation to stated performance measures and tests compliance with agreed plans. The latter include Transit New Zealand's Statement of Intent, the Land Transport Programs prepared by the local authorities, and the policies and decisions of Transfund.

The division visits the regional offices of Transit New Zealand and the Local Authorities at appropriate intervals and reviews their internal systems (including accounting and related systems) to confirm that they are being operated correctly and in conformity with the various Acts and policies of Transfund. The division carries out this work under the standards for internal auditing laid down by the New Zealand Institute of Internal Auditors. Technical and economic audits are made on a regular, planned basis about every five years, while procedural audits are made every three years. The purpose of the procedural audits is to assess the accuracy of the financial assistance claims made by the road authorities and the extent to which the road authorities are complying with Transfund's policies with regard to the custody, recording, and utilization of road fund resources.

United States: Federal Highway Trust Fund

The United States established the Highway Trust Fund in 1956 to finance the federal share of the interstate highway network and support most other federal-aid highway projects. Later amendments extended funding to other transport programs as follows:

• The Highway Safety Act of 1966 made funds available for state and community road safety programs.

• In 1982 the scope was widened to permit the financing of mass transit.

• In 1991 the Intermodal Surface Transportation Efficiency Act (ISTEA) confirmed the new role of the Highway Trust Fund as an "Intermodal Fund" by extending support to high-speed rail lines and bike trails.

The funding system involved earmarking certain road-related taxes and depositing them into a special account, or road fund. The special account was introduced primarily to finance construction of the interstate highway network and was based on the user-pay concept. The concept involves two elements: first, the user pays, and, second, the government credits the user fees directly to a highway special account to avoid confusing them with other government revenues. The user-pay concept is well established in the United States. All but six states now dedicate their user-fee revenues to special highway or transportation accounts.

The U.S. Federal Highway Trust Fund exists only as an accounting mechanism. The taxes earmarked for the Trust Fund are deposited into the general fund of the U.S. Treasury, and a paper transfer of these taxes is made to the Trust Fund as needed. Earmarked tax revenues in excess of those required to meet current expenditures are invested in public debt, and interest earned is credited to the Trust Fund. The Trust Fund finances the federal-aid highway program, administered by the Federal Highway Administration (FHWA). Since 1982 a portion of the Fund has also been used to finance mass transit projects administered by the Urban Mass Transportation Administration. Revenues from the highway portion of the Trust Fund are used to reimburse states for expenditures on approved projects. These include heavy maintenance (reconstruction, rehabilitation, and resurfacing), road improvement, new construction, road safety programs, studies, and other highway-related expenditures. The Trust Fund does not currently finance regular maintenance.

Trust Fund revenues are derived from a variety of highway user taxes, including: motor fuel taxes on gasoline, diesel, and gasohol; a graduated tax on tires weighing 40 lbs. or more; a retail tax on selected new trucks and trailers; a heavy-vehicle use tax on all trucks with a gross vehicle weight (GVW) of more than 55,000 lb.; and interest on the Trust Fund balance. Tax rates are adjusted as part of the regular budgetary process. In 1995, the tax rates were: gasoline, 12 cents/gallon; diesel, 18 cents/gallon; special fuels, 12 cents/gallon; tires, sliding incremental scale which varies from 15 cents/lb. to 50 cents/lb. over 90 lbs.; a 12 percent tax on the retail price of trucks over 33,000 lbs. GVW and trailers over 26,000 lbs. GVW; $100 plus $22 for each 1,000 lbs. over 55,000 lbs. GVW up to a flat fee of $550 for trucks over 75,000 lbs. GVW; and interest at about 6.75 percent per year. In 1995 the revenues from the above tax rates were $17.323 billion (about two-thirds from gasoline), $395 million, $2.0 billion, $682 million, and $548 million, respectively, giving a total of $20.967 billion for the year. An additional $2.8 billion was paid into the Mass Transit Account during the year.

Some vehicles, like school buses and state, local government, and nonprofit vehicles are exempted from paying federal highway motor fuel taxes. In addition, fuels purchased for off-highway uses (such as agriculture and industry) are exempt from these taxes. Off-highway uses are dealt with by coloring untaxed diesel and testing nonexempt diesel vehicles to ensure they are using regular (taxed) fuel.

The federal-aid highway program is a reimbursable program. The states are allocated a line of credit against which they can draw to meet obligations. Funds are allocated on the basis of formulas that, though not perfect, are difficult to change. The U.S. Government Accounting Office has recently criticized these formulas, but concluded that, "because the selection of a highway apportionment formula is a judgment for the Congress, GAO is making no specific recommendations." In other words, the allocation formulas are (at least in the United States) highly political. The formulas are relatively simple and generally use variables like population, road mileage, and traffic density. For example, heavy maintenance funds are allocated according to the following formula:

(interstate lane miles/total interstate miles)*0.55 +
(vehicle miles on interstate roads/total interstate vehicle miles)*0.45

This formula means that the average allocation per state is about 2 percent of total maintenance allocations, subject to each state receiving a minimum allocation of 0.5 percent. Allocations do not cover all costs, but generally cover 80 percent of costs or, in the case of maintenance, 90 percent of costs (funds from the Highway Trust Fund are provided on a cost-share basis). In looking at this allocation formula, the General Accounting Office suggested consideration of two alternatives:

Alternative 1: Distribution based equally on total lane miles and total vehicle miles traveled.

Alternative 2: Distribution based equally on total lane miles, interstate vehicle miles traveled, and state population.

Payment for work financed through the Highway Trust Fund is made in the following way:

• Work is done by a contractor.
• The contractor is paid by the state.
• Vouchers for reimbursement (usually covering several project withdrawals) are sent to FHWA for review and approval.
• Claims are certified by FHWA (this is a formality, certification is automatic).
• Certified schedules are submitted to the Treasury.
• The federal share is transferred to a state bank account by electronic transfer.

Each state participating in the scheme is required by law to carry out an annual audit. The audits are normally carried out by outside auditors and cover financial matters, compliance, and internal control procedures (that is, the audit is more extensive that a purely financial audit in that it also covers control procedures). Staff from FHWA also check these procedures on an ad hoc basis. There is no formal technical audit. Staff from FHWA used to carry out field inspections but they do not any longer because of staff shortages. However, occasional field inspections are still carried out.

FHWA is also subjected to an annual audit to ensure it follows established procedures and can account for funds spent.

About 3,000 staff manage the federal-aid highway program. They are stationed in Washington, D.C. and in each of the states.

Annex 5. Draft Road Fund Administration Bill

A Bill
entitled

An Act to make provision for the establishment of the National Road Fund Administration (the Administration) and for purposes connected therewith and incidental thereto. Enacted by the Parliament of .... as follows:

Part I—Definitions

1. This Act may be cited as the National Road Fund Administration Act, 199.., and shall come into force on such date as the Minister shall, by notice published in the Gazette, appoint.

2. In this Act, unless the context otherwise requires—
"Administration" means the National Road Fund Administration;
"Board" means the Board of Directors of the National Road Fund Administration;
"Secretary" means the Executive Secretary of the National Road Fund Administration;
"Minister" is the minister responsible for public roads......;[1]
"Public road" has the same meaning as that ascribed to it in the Public Roads Act;[2]
"Road" has the same meaning as that ascribed to it in the Public Roads Act;
"Road agency" includes any institution or body, whether or not incorporated, charged under any law with the responsibility of, or, designated as a road agency by the Minister, by notice published in the Gazette, for purposes of maintaining, rehabilitating, or developing public roads.

Part II—Establishment of Administration

3. There is hereby established a body to be known as the National Road Fund Administration of .... which shall—
   (a) be a body corporate with perpetual succession;
   (b) have a common seal;
   (c) be capable of—
      (i) acquiring, holding and disposing of real and personal property;
      (ii) suing and being sued in its corporate name; and
      (iii) doing or performing all such acts and things as a body corporate may legally do or perform.

4. The purpose of the Administration shall be to—
   (a) ensure that public roads are maintained and rehabilitated at all times;
   (b) raise funds for the maintenance and rehabilitation of public roads; and
   (c) advise the Minister on—
      (i) The preparation and the efficient and effective implementation of the annual national roads program referred to in Part VII; and
      (ii) The control of overloading of vehicles on public roads.

Part III—Board Of Directors

5. (1) The operations of the Administration shall be managed and controlled by a Board that shall consist of the following members to be appointed by the Minister—
   (a) Four ex-officio members, being nominees of each of the following Ministries:

(i) Ministry of Finance;

(ii) Ministry of Works;

(iii) Ministry of Transport and Communications; and

(iv) Ministry of Local Government.

(b) Six nongovernmental members, being nominees elected from the following constituencies:

(v) Chamber of Commerce and Industry;

(vi) Bus and Taxi Operators Association;

(vii) Road Transport Operators Association;

(viii) National Association of Tourism Operators;

(ix) Institution of Engineers; and

(x) National Farmers Association.

(c) Two other nongovernmental members, being nominees of the Board. *[Alternatively, two members, being nominees of one urban and one rural district council].*

(2) All ex officio members shall not be officers holding office below the level of Director or equivalent and shall be appointed by their respective ministers.

(3) The members of the Board shall, at the first meeting of the Board, elect a Chairperson and Vice Chairperson from among their members. The Chairperson shall be endorsed by the Minister.

(4) Members of the Board shall not, by virtue only of their appointments to the Board, be deemed to be officers in the public service.

(5) The names of all members of the Board as first constituted and every change in membership thereafter shall be published in the Gazette by the Minister.

(6) A member of the Board, other than an ex officio member, shall hold office for a period of three years from the date of his or her appointment and shall be eligible for re-appointment for one further term at the expiration of that period.

6. If a member of the Board acquires any pecuniary interest, direct or indirect, in any contract, proposed contract, or in any other matter in which private interests conflict the duties as a member and that is the subject of consideration by the Board, shall, as soon as he or she is aware of the interest in the contract, proposed contract, or any other matter, disclose such facts to the Board.

7. (1) The Board may appoint such number of committees as may be necessary for the proper discharge of the functions of the Board consisting of some members and such other persons with prescribed qualifications, and define the objectives of such groups or committees.

(2) The provisions of this Act relating to meetings of the Board shall apply *mutatis mutandis* to the meetings of the committees.

(3) The Board shall appoint the Chairperson of each committee from among the members of the board.

8. The Board may, in its discretion, at any time and for any length of period invite any person to attend any deliberations of the Board, but such person shall not be entitled to vote on any matter at any meeting of the board.

9. The office of a member, other than an ex officio member, shall be vacated-

(a) upon the expiry of the period of appointment;

(b) upon his death;

(c) if his nomination is withdrawn by the organization he represents;

(d) if he is adjudged bankrupt;

(e) if he is sentenced for an offense against any written law to a term of imprisonment of, or exceeding, six months, otherwise than as an alternative to, or in default of, the payment of a fine;

(f) if he is convicted of an offense involving fraud or dishonesty;

(g) if he has been absent from three consecutive meetings of the Board of which he has had notice without the permission of the Chairperson; or

(h) if, in the opinion of the Board, he becomes by reason of mental or physical infirmity, incapable of performing his duties as a member of the Board.

10. (1) The Board shall meet at such place and at such times as the Chairperson may determine and shall meet at least once per month.

(2) Ordinary meetings of the Board shall be convened by at least fourteen days written notice to the members by the Chairperson. The Chairperson may, at his discretion, and shall at the written request of not less than four members of the Board and within seven days of such request, convene a special meeting of the Board to transact any extraordinary business on a date

specified in the request. A written notice shall be addressed and sent to the members at least three days prior to the date of the meeting.

(3) The Chairperson or, in his absence, the Vice Chairperson shall preside at each meeting of the Board. The quorum necessary for the transaction of the business shall be five members present at any meeting of the Board.

(4) When the Chairperson and Vice Chairperson are both absent, the members present shall appoint a Chairperson to preside at the meeting.

(5) Subject to the provisions of this Act, the Board may make standing orders for the regulation of its proceedings and business or the proceedings and business of any of its committees and may vary, suspend, or revoke such standing orders.

(6) The minutes of every meeting of the Board shall be recorded in a register by the secretary of the Board and confirmed at the next succeeding ordinary meeting.

(7) The Board decisions shall be taken by the majority vote and, when the votes are equal, the Chairperson has a casting vote, with dissenting members having the right to have their views recorded in the minutes.

(8) Members of the Board shall be paid from the road fund such allowances as the Board may, subject to the approval of the Minister, determine and the Board may make provision for the reimbursement of any reasonable expenses incurred by a member of the Board or a committee of the Board in connection with the business of the Board or the committee.

Part IV—Functions and Powers of the Board

11. The functions of the Board are:
(a) to administer and manage the road fund;
(b) to ensure that all tenders for the maintenance, rehabilitation, and development of public roads are conducted through open and competitive bidding, in a transparent and fair manner;
(c) to improve arrangements for collecting road-user charges to minimize avoidance and evasion;
(d) to recommend to the Minister, from time to time, appropriate levels of road-user charges, fines, penalties, levies, or any other sums to be collected under this Act and paid into the Fund;
(e) to identify and recommend to the Minister donor funding for the maintenance, rehabilitation, and development of public roads;
(f) to establish the allocation criteria to be used to divide moneys among the various road agencies;
(g) to ensure that road agencies carry out effective monitoring of the condition of all public roads for the purpose of timely implementation of road maintenance, rehabilitation and development programs;
(h) to institute an integrated and coordinated approach to planning of road works by establishing the form and content of the Annual Road Program;
(i) to provide guidance and establish procedures to be followed in the preparation of the Annual Road Program by the various road agencies;
(j) to review and approve the Annual Road Program;
(k) to establish procedures for disbursing funds for the Annual Road Program;
(l) to prepare, publish, and submit to the Minister audited annual accounts of the Fund; and
(m) to publish periodic reports on the activities and achievements of the Administration and make the reports available to the general public.

12. Subject to the Finance and Audit Act, the Board may raise on behalf of the Administration, moneys by way of loans or bank overdrafts on such reasonable terms and conditions as the Board may in writing agree with the lender.

13. The Board shall be responsible and accountable to the Minister for ensuring efficiency, transparency and propriety in the—
(a) collection and utilization of public funds under this Act;
(b) conduct of its business; and
(c) operations and activities of the Administration.

14. *[The legislation will often include a special section here outlining the procedures to be followed and penalties that may apply if the Minister has reason to suspect that the Board has failed in its performance, has performed any act without due authority, or has willingly participated in any fraudulent activity.]*

Part V—Secretariat

15. (1) The Board will be assisted by a Secretariat headed by an Executive Secretary. The Secretariat shall be responsible for the day-to-day management of the Administration and for implementation of the decisions of the Board.

(2) The Executive Secretary shall be appointed by the Board and shall perform such functions as the Board may direct or delegate to him or her. The Executive Secretary will also act as secretary to the Board.

(3) The terms and conditions of employment of the Secretariat shall be decided by the Board based on a comparison of best practices in other similar organizations.

Part VI—Establishment of the Road Fund

16. (1) There is hereby established a fund to be known as the Road Fund.

(2) The Fund shall consist of—

(a) such road-user charges as may, from time to time, be determined by the Minister, by order published in the Gazette, on the recommendation of the Board; [or, alternatively, "such road user charges as may, from time to time, be determined by the Administration and published in the Gazette in accordance with the relevant provisions of any regulations made under section 25;"]

(b) such sums that may be appropriated by Parliament for purposes of the Fund;

(c) such sums or assets as may accrue to or vest in the Fund whether in the course of the exercise by the Board of its function or powers, or otherwise;

(d) grants, subsidies, bequests, donations, gifts, and subscriptions from Government or any other person;

(e) the sale of any property, real or personal, by or on behalf of the Administration;

(f) sums received by the Fund by way of voluntary contributions;

(g) penalties and fines imposed on overloaded vehicles; and

(h) sums that may be donated or loaned by any foreign government, international agency, or other external body of persons, corporate or undesignated.

17. (1) The purpose of the Fund shall be to finance—

(a) the administrative expenses associated with the execution of the duties and responsibilities of the Authority and the management of the Fund;

(b) routine, recurrent, and periodic maintenance of public roads;

(c) on a cost-sharing basis, the routine, recurrent, and periodic maintenance of local government roads and of undesignated roads, tracks, and trails;

(d) any monetary contribution required to be made by the Government for the implementation and execution of a donor-funded project for the maintenance, rehabilitation, or development of any public road;

(e) such road safety projects as the Board may determine;

(f) the enforcement of the limits on weights and dimensions of vehicles; and

(g) research related to the maintenance and development of roads.

(2) Any surplus from the road fund, not exceeding ..... percent of the total revenue collected or estimated to be collected in any financial year, may be utilized to finance such minor road works including upgrading of existing public roads as the Board may, on the recommendation of the road agencies, approve.

18. The Board shall ensure that in any financial year expenditures and commitments from the Fund shall not exceed the annual income of the Fund. If, however, in exceptional circumstances, the income of the Fund together with any surplus income brought forward from a previous year, is insufficient to meet the actual or estimated liabilities of the Administration, the Minister of Finance may make advances to the Fund in order to meet the deficiency or any part thereof and such advances shall be made on such terms and conditions, whether as to repayment or otherwise, as the Minister may determine, provided that such advances shall be repaid from the income of the Fund in the next financial year.

Part VII—Annual Road Program(s)

19. An Annual Road Program(s) shall be prepared at least three months before the start of the new fiscal year in such form and containing such details as may be prescribed by the Board. The Program(s) shall be prepared by the road agencies responsible for maintaining the road network or by agents designated for this purpose by the Board.

20. The Board shall review the Annual Expenditure Program(s) decide on—
    (a) the affordability of the overall program(s); and
    (b) the appropriateness of the amounts allocated for each class of road.

21. The Board shall transmit to the Minister of Public Works and Housing and the Directors of other road agencies together with the Minister of Finance the approved Annual Road Program(s).

22. Pursuant to Section 11, the Board may recommend a rise in the level of the road tariff to ensure it generates sufficient revenues to finance the approved Annual Road Program(s) and shall provide the Minister with an estimate of the additional income to the road fund from such increases.

Part VIII—Accounts

23. (1) The Board shall cause to be kept proper books and other records of account in respect of receipts and expenditures of the road fund in accordance with acceptable principles of accounting.

(2) The accounts of the road fund shall be audited annually by independent professional auditors nominated by the Board and approved by the Auditor General's Office. The expenses of the audit shall be paid out of the road fund.

(3) The auditors shall complete their audit of the accounts within three months of the end of each financial year and shall include in their report assessments relating to the achievement of the objectives of the Administration;compliance with the policies, procedures,and criteria established by the Board; and the effectiveness of the management of the road fund.

(4) The Board shall, as soon as is practicable, but not later than six months after the end of the financial year of the Administration, submit to the Minister an annual report on all the financial transactions of the road fund and on the work, activities, and operations of the Administration.

(5) The Authority shall at all times comply with the provisions of the Finance and Audit Act.

24. (1) All sums received for the purposes of the Administration shall be paid into a banking account, and no amount shall be withdrawn therefrom except under the authority of the Board and by means of checks signed by such persons as are authorized in that behalf by the Board.

(2) Any part of the road fund not immediately required for the purposes of the Administration may, on the recommendation of the Board, be invested in such manner as the Board may, in its discretion, determine. *[The legislation may limit these investments to bills carrying a Standard and Poor's or Moody's rating of "A" or better.]*

(3) The financial year of the Authority and the Fund shall be the period of twelve months commencing on the 1st of April of each year and ending on the 31st March of the following year. The first financial year may be shorter of longer than twelve months as the Board may determine, but in any case not longer then eighteen months.

Part IX—Miscellaneous

25. The Minister shall, by notice published in the Gazette, make regulations stipulating the detailed procedures to be followed by the Board regarding the works to be financed through the road fund, procedures to be followed in preparing the Annual Road Program(s), procedures for allocating funds among the different road agencies, arrangements for disbursing funds for road works, and the detailed financial management procedures to be followed by the Board.

Draft Regulations for the Management of the Road Maintenance Fund

National Road Fund Administration Regulations 199..

under s.25

1. These regulations may be cited as the National Road Fund Administration Regulations 199.. and shall come into operation on the date of publication in the Gazette.

under s.1

2. Within 14 days from the date of Presidential assent to the National Road Fund Administration Bill, the Minister shall cause a notice to be published in the Gazette appointing the effective date of the Act.

under s.5

3. Within 28 days from the effective date of the Act, the Minister, in consultation with the various sectors required to be represented on the Board of the Authority, shall appoint members of the Board.
4. Within 14 days after expiry of the 28 days in Regulation 3, the Minister shall cause a notice of the members appointed to the Board to be published in the Gazette, specifying the place, date, and hour of the first meeting of the Board.

under s.7

5. The Board may delegate any of its powers to committees consisting of such member or members of its body as it may consider fit, or expedient, and any committee so formed shall conform to any regulations or direction of the Board.
6. The Board and its subcommittees may appoint such study groups or committees as may be necessary for the proper discharge of its functions consisting of some members and other persons with such prescribed qualifications as may be required, and define the objectives of such groups or committees.
7. The Board and its subcommittees may co-opt any person to advise it during its deliberations, provided that any person so co-opted shall not be entitled to vote at any meeting of the Board or of its subcommittees.

8. For the better performance of its functions, the Board and its subcommittees shall, subject to the provisions of the Act, have power to:

(a) affiliate or cooperate with government departments; universities; technical colleges; persons engaged in the maintenance, rehabilitation, or development of public roads; and such other organizations or persons as may appear to the Board to be proper or beneficial to associate with; and

(b) publish from time to time such technical and other information as it deems necessary or expedient for the promotion of knowledge on the maintenance, rehabilitation, and development of public roads.

under s.11

9. The Road Fund shall be managed by the Board who shall:

(a) devise and put in place a mechanism for collecting road-user charges;

(b) when relevant, devise and put in place arrangements for collecting from the Treasury any road-user charges collected for the Road Fund;

(c) establish and publish the criteria to be used to divide Road Fund revenues among the different road agencies entitled to draw on the Road Fund, where such criteria may be based on the condition of the road network, the type of maintenance required (whether routine or periodic), the length of the road network, and the volume of traffic;

(d) negotiate an annual Framework Agreement with the Ministry of Finance establishing the procedures to be followed when adjusting the road-user charges during the year concerned, which shall include the general financial policies of the Administration, the maximum annual increase in the road-user charges, the size of the Administration's administrative budget, and any matters which that have an impact on the Government's fiscal and macroeconomic policies;

(e) establish procedures for disbursing funds for works forming part of the approved Annual Road Program(s);

(f) establish and publish procedures that ensure that non-transport users of diesel are not unduly penalized by introduction of the road maintenance levy; and

(g) advise the Minister on ways to control the overloading of vehicles, particularly on international transit routes.

10. Major upgrading and new works will continue to be financed through the government's development budget, and all financial resources made available for such purpose shall be channeled through the Road Fund.

under s.15

11. The Secretariat will consist of no more than [....] regular staff who shall be appointed by the Board on the recommendation of the Executive Secretary. A firm of chartered accountants, or a bank, may be appointed to act as Secretariat or to assist the Secretariat.

12. Without prejudice to the generality of these regulations, the Secretariat shall be responsible for:

(a) keeping proper accounts and records in respect of the Fund;

(b) maintaining separate bank accounts for local and donor funds, in which shall be recorded all receipts into the Fund and all disbursements from the Fund;

(c) preparing and submitting for audit in respect of each financial year a balance sheet, a statement of income and expenditure, and a statement of cash flow in such forms and manners as the Administration may prescribe;

(d) preparing the Annual Report of the Fund in such form and with such content as may be prescribed by the Authority; and

(e) arranging the business for meetings of the Board and its subcommittees.

13. The Administration shall, at such intervals as the Minister may, by order in writing, require, submit to the Minister reports and financial statements in such form as the Minister may by like order determine, regarding the operations and activities of the Administration and the Fund.

under s.16

14. The road user charges referred to in section 16 (2) (a) shall consist of

(a) a surcharge on the price of gasoline and diesel fuel to be known as the fuel levy. The said fuel levy shall be a charge over and above ordinary import duties, general sales taxes, and other charges on fuels, and shall be used exclusively as a source of revenue for the Road Fund;

(b) international transit charges to be paid by foreign vehicle operators using the roads of ....; and

(c) vehicle license fees.

The road user charges mentioned in paragraphs (a) to (c) above shall be subject to revision by the Minister from time to time on the recommendation of the Board, and upon such revision the public shall be duly informed of the same through the press.

15. The road-user charges shall, to the extent possible, be collected under contract, and the proceeds shall be directly deposited into the Administration's bank accounts. Contracts will be entered into with the Oil Companies, the Department of Customs, the Ministry of Transport and Communications, and/or with private contractors. Otherwise, the collection shall be the responsibility of the Treasury, provided that having been so collected the moneys shall without delay be transferred by the Treasury to the Road Fund.

16. The Administration shall open and maintain separate bank accounts for each of the sources of funds allocated to the Road Fund.

17. All moneys provided by international donors for the Road Fund shall be given by the donor directly to the Fund and not through the Government.

18. Disbursement of moneys from any of the accounts holding donor funds shall be subject to the provisions of section 11 and the prior authority of the relevant donor.

under s.17

19. The detailed basis of the cost-sharing arrangements will be decided by the Administration and shall be published and revised from time to time.

20. Urban and Rural District Councils shall use revenue from rates, local taxes, and other local revenue sources to contribute to the financing of routine and periodic maintenance of roads, tracks, and trails under their responsibility.

21. Individuals and communities living in areas with unclassified roads shall also be entitled to receive funds for maintenance on a cost-share basis. Such groups shall be required to register their interests in the roads, form themselves into local roads committees, and agree on cost-sharing arrangements for maintaining their roads.

22. Local roads committees may contribute their share of the costs in the form of materials, direct labor, cash or by a combination of any or all of these.

23. Funds shall be disbursed only for goods and services forming part of the approved Annual Road Program(s) according to procedures to be established by the Board.

24. Work undertaken by contractors with a value in excess of $ ... shall be certified by a registered engineer, and, once so certified, payment shall be made directly to the contractor.

25. Work undertaken by small-scale contractors, or by force account, will be subject to similar controls to be agreed among the Administration, the Minister, and the various District Councils.

under s.19

26. The Annual Road Program(s) shall allocate the revenues of the road maintenance fund to various categories of roads for the year, following the allocation criteria prescribed by the Board. Without prejudice to the other factors that the Board may take into account in determining the allocation criteria, some of the major factors to consider shall be the condition of the road network, the type of maintenance required (whether routine, recurrent, or resurfacing), the length of the road network, and the volume of traffic.

27. Funds shall be withdrawn from the Road Fund on presentation of a check signed by two authorized signatories, being either the Chairperson of the Board and the Secretary, or one of them and a designated member of the Secretariat.

28. In the interim, and until such time as the road agencies have developed the capacity to prepare their submissions to the Administration, the Administration may enter into a contract with local consultants for the purposes of assisting the road agencies to prepare their individual road programs.

29. The consultants so appointed shall work in close consultation with the concerned central and local government agencies to assist the various road agencies with the preparation of their road maintenance, rehabilitation, and development programs; set priorities; and consolidate the individual programs into an overall Annual Road Program to fit within the available resources. Such plans shall include medium-term maintenance programs and longer-term rehabilitation and development programs.

Notes

1. It could also be the Minister of Transport, Minister of Finance, or Prime Minister's Office.

2. The Act may need to be amended if the road fund intends to finance undesignated (community) roads.

Annex 6. Standard Format for Setting Up a Road Fund Under Existing Legislation

Financing (Roads Fund) Regulations 199?

In exercise of the powers conferred on me by section ... of the Finance Order 19...., *[or other relevant Orders or Decrees]*

....

Minister of Finance and Economic Planning *[adjust as needed]* make the following Regulations:

Citation and Commencement

1. These regulations may be cited as the Finance (Roads Fund) Regulations 199? and shall come into operation on the date of publication in the Gazette.

Interpretation

2. In these regulations, unless the context otherwise requires,

"Appointed member" means a member of the Board who is appointed by the Minister under regulation 9;

"Board" means the board constituted under regulation 8;

"Secretary" means the Executive Secretary of the Fund appointed under regulation 12;

"Fund" means the Roads Fund established by the *[relevant Order or Decree]*, 199?; and

"Minister" means the minister responsible for Finance.

Purpose of the Fund

3. The purpose of the Fund is to finance

(a) routine and periodic maintenance of all classified roads under the jurisdiction of the Ministry of Works and of the Ministry of Local Government *[adjust as needed]*;

(b) on a cost share basis, Urban Council roads and the unclassified roads under the jurisdiction of Development Councils;

(c) road safety projects; and

(d) a limited amount of road upgrading, rehabilitation, and new works.

Road Fund Revenues

4. The Road Fund shall have one or more commercial bank accounts into which the following fees and charges (the road tariff) shall be deposited:

(a) vehicle license fees, including any supplementary heavy vehicle fees that may be introduced by the Road Fund Board;

(b) a road maintenance levy on petrol and diesel;

(c) fines imposed on overloaded vehicles; and

(d) any other road-user charges and/or donor funding that may from time to time be allocated by Parliament.

These revenues constitute the road tariff and no one will be exempted from paying it.

Collection and Deposit Procedures

5. The Board will undertake all necessary actions to ensure that:

(a) license fees and any heavy vehicle license fees that may be introduced by the Board are separately deposited into the Consolidated Fund by the Department of Customs and the [vehicle licensing authority], and thereafter directly deposited into the Road Fund bank account; and

(b) the road maintenance levy on petrol and diesel is separately deposited into the Consolidated Fund by the oil companies and thereafter directly deposited into the Road Fund bank account.

The Board will take steps to ensure that all funds due

to the Road Fund are collected and deposited in a timely manner into the Road Fund bank account.

Authorized Expenditures

6. The Road Fund shall be used primarily to finance routine and periodic maintenance, which shall remain as the first charge on the Road Fund. The Road Fund will also meet the costs of administering the Road Fund. Once all road maintenance requirements have been met, the remaining funds shall be used exclusively to finance selected road safety projects, road rehabilitation, minor improvements, and new works (minor improvement and new works not to exceed ...... percent of annual revenues).

7. Roads under the jurisdiction of the Ministry of .... *[that is, the trunk road network]* will be fully funded by the Road Fund, while roads under the jurisdiction of the municipalities *[or other local government agencies]* will be financed on a cost-sharing basis. The detailed basis of the cost-sharing arrangements will be decided by the Board, published, and revised from time to time. Municipalities will be expected to contribute their share of the costs using revenues from rates and other local taxes. Individuals and communities living in areas with unclassified roads will also be entitled to receive funds for maintenance. Such groups will first have to register their interests in these roads, form themselves into local roads committees, and agree on cost-sharing arrangements for maintaining these roads. Local roads committees may contribute their share of the costs in the form of materials, direct labor, and/or cash.

Management of the Road Fund

8. The Road Fund will be managed by a National Roads Board, which will report to the Ministry of .... as its parent ministry. The Board will manage the Road Fund in an executive capacity and advise the Minister on all matters pertaining to the financing of roads. Among other things, the Board will:

(a) improve arrangements for collecting all the fees and charges assigned to the Road Fund to minimize avoidance and evasion;

(b) institute an integrated and coordinated approach to the planning of road works by establishing the form and content of the Annual Road Program;

(c) establish and publish the criteria used to divide Road Fund revenues among the different road agencies entitled to draw on the Road Fund;

(d) review and approve the Annual Road Expenditure Program prepared by the various implementing agencies;

(e) recommend to the Minister of Finance the level of fees and charges required to finance the recommended road maintenance program for inclusion in the government's annual or supplementary budget;

(f) mobilize a publicity program to inform the public about the maintenance programs being financed from the Road Fund, assure the public that the Road Fund is well managed, and seek their support for possible increases in the level of the road user charges as and when such increases are needed; and

(g) establish procedures for disbursing funds for works forming part of the approved Annual Expenditure Program.

Composition of the National Roads Board

9. The Board will be appointed by the Minister of .... and will consist of twelve members: the Chairperson, four ex officio members representing government departments, five members representing nongovernmental organizations, and two members representing municipalities *[or other local government agencies]*. The members of the Board will be as follows:

(a) (1) The Chairperson of the Board;

(b) Four ex officio members, being nominees of the following Ministries:

(2) Ministry of Finance;

(3) Ministry of Transport;

(4) Ministry of Local Government; and

(5) Ministry of Energy.

(c) five nongovernmental members, being nominees of the following organizations:

(6) Chamber of Commerce and Industry;

(7) Bus and Taxi Operators Association;

(8) Road Transport Operators Association;

(9) Association of Consulting Engineers (or Institution of Engineers); and

(10) National Farmers Association.

(d) (11) two members, being nominees of municipalities *[or other local government agencies]*.

10. The Chairperson of the Board will be appointed by the Minister of ...., following consultations with

the Board. The ex officio members will not be below the level of Director, or the equivalent. Members of the Board will be appointed for a term of two years. Members of the Board shall cease to be members if their nomination is canceled by the organization responsible for nominating them.

11. If a member of the Board acquires any pecuniary interest, direct or indirect, in any contract or proposed contract being considered by the Board, or in any other matter in which his private interests conflict with his duties as a member of the Board, he shall, as soon as he becomes aware of his interest in the contract or proposed contract or any other matter, disclose the facts to the Board and withdraw from all meetings at which such matters may be discussed.

12. The Board may establish subcommittees dealing with subjects like: road safety, environment, engineering, and road fees. The Board may also invite additional nonvoting members to attend any of its meetings.

Meetings of the Board

13. The Board shall meet at least once a month for a regular board meeting at a time and place decided by the Chairperson. The Chairperson shall, at the written request of not less than four members of the Board, convene a special meeting of the Board to transact any extraordinary business on a date specified in the request. A written notice of such a meeting shall be sent to the members at least three days prior to the date of the meeting.

14. At all meetings of the Board the quorum necessary for the transaction of the business shall be a majority of the members then in office. The Board decisions will be taken by majority vote, and when the votes are equal, the Chairperson shall have a casting vote, with the dissenting members having the right to have their views recorded in the minutes.

15. The minutes of every meeting of the Board shall be recorded in a register by the Secretary of the Board and signed by the Chairperson of the meeting and the Secretary.

16. Members of the Board will be compensated for the time spent attending board meetings.

Road Fund Secretariat

17. The Board will be assisted by a Secretariat headed by an Executive Secretary. The Secretariat will be responsible for the day-to-day management of the Road Fund and for implementing the decisions of the Board. The Executive Secretary will be appointed by the Board and shall perform such functions as the Board may direct or delegate to him. The Executive Secretary will also act as Secretary to the Board.

18. The Secretariat will consist of no more than [...] staff. A firm of chartered accountants, or a bank, may be appointed to act as Secretariat.

19. Among other things, the Secretariat will be expected to:

(a) keep proper accounts and records in respect of the Road Fund;

(b) maintain the Road Fund bank account in which shall be recorded all receipts into the Fund and all disbursements from the Fund;

(c) prepare monthly statements of revenues collected, amounts deposited into the Road Fund bank accounts, commitments entered into by the Board, withdrawals authorized, and actual withdrawals;

(d) prepare and submit for audit in respect of each financial year a statement of income and expenditure, a statement of cash flow, and such other financial statements as the Accountant General may prescribe;

(e) prepare the Annual Report in such form and with such content as prescribed by the Board; and

(f) prepare the Agenda and arrange the meetings of the Board.

Annual Road Program

20. At least three months before the beginning of each fiscal year, the Board shall review the Annual Road Program for that year. The Annual Road Program, in such form and containing such details as may be prescribed by the Board, shall be prepared by the road agencies responsible for maintaining the road networks funded by the Road Fund.

21. The Annual Road Program shall comprise:

(a) the Annual Expenditure Program for the next year; and

(b) the revenue projections of the Road Fund for the next year.

22. The Annual Expenditure Program shall allocate the revenues of the Road Fund to various categories of roads for the year, following the allocation criteria prescribed

by the Board. Allocation criteria may be based on the condition of the road network, the type of maintenance required (routine or periodic), the length of the road network, the volumes of traffic, the population served by the roads, and any other factors decided by the Board.

23. In consultation with the Minister of Finance, the Board will review the Annual Expenditure Program to be financed by the Road Fund and decide on the affordability of the overall program and the appropriateness of the amounts allocated for each class of road.

24. The Board shall transmit to the Minister of .... and to the Minister of Finance the approved Annual Expenditure Program.

25. Pursuant to regulation 8 (e), the Board may recommend to the Minister of Finance any increase in the level of fees and charges required to finance the approved Annual Expenditure Program, and will provide an estimate of the additional income to the Road Fund from such increases.

Disbursement of Funds

26. Funds will be disbursed only for goods and services forming part of the approved Annual Expenditure Program and according to procedures to be established by the Board. Work undertaken by contractors with a value over $ ... must be certified by a registered engineer, and payment will then be made directly to the contractor. Work undertaken by small-scale contractors or using in-house staff and equipment will be subject to similar controls to be agreed between the Board and the Minister of ....

Withdrawal Procedures

27. Funds will be withdrawn from the Road Fund on presentation of a check signed by two authorized signatories: either one member of the Board and the Executive Secretary or one member of the Board and a designated Accountant from the Ministry of Finance. *[other options are also possible]*

Audits

28. The accounts and other financial statements of the Road Fund will be audited annually by an independent firm of auditors selected by the Auditor General. The auditor will be expected to use international audit standards. The auditor will present a report to the Board that will give an opinion on the accuracy of the records and financial accounts of the Road Fund, the completeness of income of the Road Fund, the conformity of payments with the priorities laid down in regulation 6, whether disbursements are in accordance with regulation 26, and the accuracy of accounting procedures and internal control procedures.

29. Technical audits of works will also be carried out on a selective basis as recommended by the Board in consultation with the Minister of ......

Annual Report

30. Within four months after the end of each financial year, the Board will publish an Annual Report. The Annual Report will summarize the policies of the Board, the main activities of the Road Fund during the preceding year, the audited accounts for the year just ended, and the auditors report on the accounts.

[The regulations may also want to deal with: the Board's power to lend and borrow funds and the preparation of an annual contract plan between the Board and the parent ministry covering the business objectives to be followed by the Board during the ensuing year].

References

Austroads. 1996. *National Performance Indicators*. Sydney.

Bahl, Roy. 1992. "The Administration of Road User Taxes in Developing Countries." Working Paper Series 986, World Bank, Washington, D.C.

Churchill, A. 1972. *Road User Charges in Central America*. Baltimore: Johns Hopkins University Press.

Creightney, Cavelle. 1993a. "Road User Taxation in Selected OECD Countries." SSATP Working Paper 3, World Bank, Africa Technical Department, Washington, D.C.

———. 1993b. *Transport and Economic Performance: A Survey of Developing Countries*. Technical Paper 232. Washington, D.C.: World Bank.

de Richecour, Anne B. and Ian G. Heggie. 1995. "Review of African Road Funds: What Works and Why?" SSATP Working Paper 14, World Bank, Africa Technical Department, Washington D.C.

Eklund, P. 1967. "Earmarking of Taxes for Highways in Developing Countries." Working Paper 1, World Bank, Economics Department, Washington, D.C.

ESCAP (Economic and Social Commission for Asia and the Pacific). 1995. "Review of Developments in Transport, Communications, and Tourism in the ESCAP Region." Report ST/ESCAP/1620. Bangkok.

Harral, C., and A. Faiz. 1988. *Road Deterioration in Developing Countries*. Washington, D.C.: World Bank.

Hau, T. 1992. "Congestion Charging Mechanisms for Roads: An Evaluation of Current Practice." Working Paper Series 1071, World Bank, Infrastructure and Urban Development Department, Washington, D.C.

Heggie, Ian G. 1991a. *Designing Major Policy Reform: Lessons form the Transport Sector*. Discussion Paper 115. Washington, D.C.: World Bank.

———. 1991b. "Improving Management and Changing Policies for Roads: An Agenda for Reform." Report INU 29, World Bank, Infrastructure and Urban Development Department, Washington, D.C.

———. 1992. "Selecting Appropriate Instruments for Charging Road Users." Report INU 95, World Bank, Infrastructure and Urban Development Department, Washington, D.C.

———. 1995a. "Commercializing Africa's Roads: Transforming the Role of the Public Sector." *Transport Reviews* 15(2): 167-84.

———. 1995b. *Management and Finance of Roads*. Technical Paper 275. Washington, D.C.: World Bank.

Hungary Ministry of Transport. 1996. *Value of National Public Roads*. Department of Public Roads, Communication and Water Management. Budapest.

Indian Ministry of Surface Transport. 1996. Unpublished country paper presented at ESCAP/World Bank conference on "Management and Financing of Roads in the ESCAP Region," Bangkok, September.

International Road Federation. 1996. "Paying for Roads—World Trends in Road Network Financing: Private Sector Takes the Lead." *World Highways* (Nov./Dec.): 25–26.

———. 1997. *World Road Statistics*. Geneva; Washington, D.C.

Isotalo, Jukka. 1995. *Development of Good Governance in the Road Sector in Finland*. Report 16/1995. Helsinki: Finnish Road Administration.

Kranton, Rachel. 1990. "Pricing, Cost Recovery, and Production Efficiency in Transport." Working Paper Series 445, World Bank, Infrastructure and Urban Development Department, Washington, D.C.

Local Authority Associates. 1989. *Highway Maintenance: A Code of Good Practice*. Association of County Councils. London.

Malmberg Calvo, C. 1997 "The Institutional and Financial Framework of Rural Transport Infrastructure." SSATP Working Paper 17, World Bank, Africa Technical Department, Washington, D.C.

McCleary, W. 1991. "The Earmarking of Government Revenue: A Review of Some World Bank Experience." *The World Bank Research Observer* 6(1): 81–104.

Newbery, David M., G.A. Hughes, W.D.O. Paterson, and E. Bennathan. 1988. *Road Transport Taxation in Developing Countries*. Discussion Paper 26, World Bank, Washington, D.C.

OECD (Organisation for Economic Co-operation and Development). 1994. *Road Maintenance and Rehabilitation: Funding and Allocation Strategies.* Paris.

Oum, Tae H., W.G. Waters, and Jong Song Yong. 1990. "A Survey of Recent Estimates of Price Elasticity of Demand for Transport." Working Paper Series 359, World Bank, Washington, D.C.

Robinson, Richard. 1988. *A View of Road Maintenance Economics, Policy and Management in Developing Countries.* Research Report 145. Crowthorne: Transport and Road Research Laboratory.

Robinson, Richard, and P.H. May. 1997. *Road Management Systems: Guidelines for their Specification and Selection.* Proceedings of the Institution of Civil Engineers, Transport. 123 (February): 9–16.

Schliessler, Andreas, and A. Bull. 1993. *Roads: A New Approach for Road Network Management and Conservation.* Santiago: United Nations Economic Commission for Latin America (ECLAC).

Scholss, M. 1993. "Does Petroleum Procurement and Trade Matter? The Case of Sub-Saharan Africa." *Finance and Development* 30 (March): 44–46.

Shirley, Mary. 1989. *The Reform of State-Owned Enterprises: Lessons From World Bank Lending.* Policy and Research Series 4. Washington, D.C.: World Bank.

Shirley, Mary, and John Nellis. 1991. *Public Enterprise Reform: The Lessons From Experience.* EDI Development Studies. Washington, D.C.: World Bank.

Snaith, M.S., Richard Robinson, and U. Danielson. 1997. *Road Maintenance Management.* Basingstoke: MacMillan.

South Africa Department of Transport. 1996. *A Future for Roads in South Africa.* Pretoria.

Stock, Elizabeth. 1996. "The Problems Facing Labor-Based Road Programs and What To Do About Them: Evidence from Ghana." SSATP Working Paper 24, World Bank, Africa Technical Department, Washington, D.C.

Stock, Elizabeth, and Jan de Veen. 1996. *Expanding Labor-based Methods for Road Works in Africa.* Technical Paper 347. Washington, D.C.: World Bank.

Thriscutt, S., and M. Mason. 1991. "Road Deterioration in Sub-Saharan Africa." In *The Road Maintenance Initiative: Building Capacity for Policy Reform.* Volume II. Washington, D.C.: World Bank.

TRRL (Transport and Road Research Laboratory) Overseas Unit. 1988. *A Guide to Geometric Design.* Overseas Road Note 6. Crowthorne.

———. 1997. *Guidelines for Design and Operation of Road Management Systems.* Overseas Road Note 15. Crowthorne.

UNCTAD and WTO (United Nations Conference on Trade and Development and World Trade Organization). 1996. *Applying ISO 9000 Quality Management Systems.* Geneva.

Walters, A. 1968. *The Economics of Road User Charges.* Baltimore: Johns Hopkins University Press.

World Bank. 1991a. "FY91 Transport Sector Review." Report INU-OR8, Infrastructure and Urban Development Department, Washington, D.C.

———. 1991b. *Lessons of Tax Reform.* Washington, D.C.

———. 1991c. *The Reform of Public Sector Management: Lessons From Experience.* Policy and Research Series 18, Country Economics Department. Washington, D.C.

———. 1992. "Brazil: Energy Pricing and Investment Study." Report 8502, Latin America and the Caribbean Country Department, Washington, D.C.

———. 1994. *World Development Report: Infrastructure for Development.* New York: Oxford University Press.

———. 1996a. "Bangladesh, Government That Works: Reforming the Public Sector." South Asia Division, Washington, D.C.

———. 1996b. "Morocco Impact Evaluation Report: Socioeconomic Influence of Rural Roads." Report on Fourth Highway Project, Operations Evaluation Department, Washington, D.C.

———. 1996c. "Road Maintenance by Contract: Dissemination of Good Practice in Latin America." Report 15627-LAC, Central America and Mexico Department, Washington, D.C.

Distributors of World Bank Publications

Prices and credit terms vary from country to country. Consult your local distributor before placing an order.

ARGENTINA
Oficina del Libro Internacional
Av. Cordoba 1877
1120 Buenos Aires
Tel: (54 1) 815-8156
Fax: (54 1) 815-8354
E-mail: olilibro@satlink.com

AUSTRALIA, FIJI, PAPUA NEW GUINEA, SOLOMON ISLANDS, VANUATU, AND WESTERN SAMOA
D.A. Information Services
648 Whitehorse Road
Mitcham 3132
Victoria
Tel: (61) 3 9210 7777
Fax: (61) 3 9210 7788
E-mail: service@dadirect.com.au
URL: http://www.dadirect.com.au

AUSTRIA
Gerold and Co.
Weihburggasse 26
A-1011 Wien
Tel: (43 1) 512-47-31-0
Fax: (43 1) 512-47-31-29
URL: http://www.gerold.co/at.online

BANGLADESH
Micro Industries Development Assistance Society (MIDAS)
House 5, Road 16
Dhanmondi R/Area
Dhaka 1209
Tel: (880 2) 326427
Fax: (880 2) 811188

BELGIUM
Jean De Lannoy
Av. du Roi 202
1060 Brussels
Tel: (32 2) 538-5169
Fax: (32 2) 538-0841

BRAZIL
Publicações Tecnicas Internacionais Ltda.
Rua Peixoto Gomide, 209
01409 Sao Paulo, SP
Tel: (55 11) 259-6644
Fax: (55 11) 258-6990
E-mail: postmaster@pti.uol.br
URL: http://www.uol.br

CANADA
Renouf Publishing Co. Ltd.
5369 Canotek Road
Ottawa, Ontario K1J 9J3
Tel: (613) 745-2665
Fax: (613) 745-7660
E-mail: order.dept@renoufbooks.com
URL: http://www.renoufbooks.com

CHINA
China Financial & Economic Publishing House
8, Da Fo Si Dong Jie
Beijing
Tel: (86 10) 6333-8257
Fax: (86 10) 6401-7365

China Book Import Centre
P.O. Box 2825
Beijing

COLOMBIA
Infoenlace Ltda.
Carrera 6 No. 51-21
Apartado Aereo 34270
Santafé de Bogotá, D.C.
Tel: (57 1) 285-2798
Fax: (57 1) 285-2798

COTE D'IVOIRE
Center d'Edition et de Diffusion Africaines (CEDA)
04 B.P. 541
Abidjan 04
Tel: (225) 24 6510;24 6511
Fax: (225) 25 0567

CYPRUS
Center for Applied Research
Cyprus College
6, Diogenes Street, Engomi
P.O. Box 2006
Nicosia
Tel: (357 2) 44-1730
Fax: (357 2) 46-2051

CZECH REPUBLIC
National Information Center
prodejna, Konviktska 5
CS – 113 57 Prague 1
Tel: (42 2) 2422-9433
Fax: (42 2) 2422-1484
URL: http://www.nis.cz/

DENMARK
SamfundsLitteratur
Rosenoerns Allé 11
DK-1970 Frederiksberg C
Tel: (45 31) 351942
Fax: (45 31) 357822

ECUADOR
Libri Mundi
Libreria Internacional
P.O. Box 17-01-3029
Juan Leon Mera 851
Quito
Tel: (593 2) 521-606; (593 2) 544-185
Fax: (593 2) 504-209
E-mail: librimu1@librimundi.com.ec
E-mail: librimu2@librimundi.com.ec

EGYPT, ARAB REPUBLIC OF
Al Ahram Distribution Agency
Al Galaa Street
Cairo
Tel: (20 2) 578-6083
Fax: (20 2) 578-6833

The Middle East Observer
41, Sherif Street
Cairo
Tel: (20 2) 393-9732
Fax: (20 2) 393-9732

FINLAND
Akateeminen Kirjakauppa
P.O. Box 128
FIN-00101 Helsinki
Tel: (358 0) 121 4418
Fax: (358 0) 121-4435
E-mail: akatilaus@stockmann.fi
URL: http://www.akateeminen.com/

FRANCE
World Bank Publications
66, avenue d'Iéna
75116 Paris
Tel: (33 1) 40-69-30-56/57
Fax: (33 1) 40-69-30-68

GERMANY
UNO-Verlag
Poppelsdorfer Allee 55
53115 Bonn
Tel: (49 228) 949020
Fax: (49 228) 217492
URL: http://www.uno-verlag.de
E-mail: unoverlag@aol.com

GHANA
Epp Books Services
P.O. Box 44
TUC
Accra

GREECE
Papasotiriou S.A.
35, Stournara Str.
106 82 Athens
Tel: (30 1) 364-1826
Fax: (30 1) 364-8254

HAITI
Culture Diffusion
5, Rue Capois
C.P. 257
Port-au-Prince
Tel: (509) 23 9260
Fax: (509) 23 4858

HONG KONG, MACAO
Asia 2000 Ltd.
Sales & Circulation Department
Seabird House, unit 1101-02
22-28 Wyndham Street, Central
Hong Kong
Tel: (852) 2530-1409
Fax: (852) 2526-1107
E-mail: sales@asia2000.com.hk
URL: http://www.asia2000.com.hk

HUNGARY
Euro Info Service
Margitszgeti Europa Haz
H-1138 Budapest
Tel: (36 1) 350 80 24, 350 80 25
Fax: (36 1) 350 90 32
E-mail: euroinfo@mail.matav.hu

INDIA
Allied Publishers Ltd.
751 Mount Road
Madras - 600 002
Tel: (91 44) 852-3938
Fax: (91 44) 852-0649

INDONESIA
Pt. Indira Limited
Jalan Borobudur 20
P.O. Box 181
Jakarta 10320
Tel: (62 21) 390-4290
Fax: (62 21) 390-4289

IRAN
Ketab Sara Co. Publishers
Khaled Estamboli Ave., 6th Street
Delafrooz Alley No. 8
P.O. Box 15745-733
Tehran 15117
Tel: (98 21) 8717819; 8716104
Fax: (98 21) 8712479
E-mail: ketab-sara@neda.net.ir

Kowkab Publishers
P.O. Box 19575-511
Tehran
Tel: (98 21) 258-3723
Fax: (98 21) 258-3723

IRELAND
Government Supplies Agency
Oifig an tSoláthair
4-5 Harcourt Road
Dublin 2
Tel: (353 1) 661-3111
Fax: (353 1) 475-2670

ISRAEL
Yozmot Literature Ltd.
P.O. Box 56055
3 Yohanan Hasandlar Street
Tel Aviv 61560
Tel: (972 3) 5285-397
Fax: (972 3) 5285-397

R.O.Y. International
PO Box 13056
Tel Aviv 61130
Tel: (972 3) 5461423
Fax: (972 3) 5461442
E-mail: royil@netvision.net.il

Palestinian Authority/Middle East
Index Information Services
P.O.B. 19502 Jerusalem
Tel: (972 2) 6271219
Fax: (972 2) 6271634

ITALY
Licosa Commissionaria Sansoni SPA
Via Duca Di Calabria, 1/1
Casella Postale 552
50125 Firenze
Tel: (55) 645-415
Fax: (55) 641-257
E-mail: licosa@ftbcc.it
URL: http://www.ftbcc.it/licosa

JAMAICA
Ian Randle Publishers Ltd.
206 Old Hope Road, Kingston 6
Tel: 876-927-2085
Fax: 876-977-0243
E-mail: irpl@colis.com

JAPAN
Eastern Book Service
3-13 Hongo 3-chome, Bunkyo-ku
Tokyo 113
Tel: (81 3) 3818-0861
Fax: (81 3) 3818-0864
E-mail: orders@svt-ebs.co.jp
URL: http://www.bekkoame.or.jp/~svt-ebs

KENYA
Africa Book Service (E.A.) Ltd.
Quaran House, Mfangano Street
P.O. Box 45245
Nairobi
Tel: (254 2) 223 641
Fax: (254 2) 330 272

KOREA, REPUBLIC OF
Daejon Trading Co. Ltd.
P.O. Box 34, Youida, 706 Seoun Bldg
44-6 Youido-Dong, Yeongchengo-Ku
Seoul
Tel: (82 2) 785-1631/4
Fax: (82 2) 784-0315

MALAYSIA
University of Malaya Cooperative Bookshop, Limited
P.O. Box 1127
Jalan Pantai Baru
59700 Kuala Lumpur
Tel: (60 3) 756-5000
Fax: (60 3) 755-4424
E-mail: umkoop@tm.net.my

MEXICO
INFOTEC
Av. San Fernando No. 37
Col. Toriello Guerra
14050 Mexico, D.F.
Tel: (52 5) 624-2800
Fax: (52 5) 624-2822
E-mail: infotec@rtn.net.mx
URL: http://rtn.net.mx

Mundi-Prensa Mexico S.A. de C.V.
c/Rio Panuco, 141-Colonia Cuauhtemoc
06500 Mexico, D.F.
Tel: (52 5) 533-5658
Fax: (52 5) 514-6799

NEPAL
Everest Media International Services (P) Ltd.
GPO Box 5443
Kathmandu
Tel: (977 1) 472 152
Fax: (977 1) 224 431

NETHERLANDS
De Lindeboom/InOr-Publikaties
P.O. Box 202, 7480 AE Haaksbergen
Tel: (31 53) 574-0004
Fax: (31 53) 572-9296
E-mail: lindeboo@worldonline.nl
URL: http://www.worldonline.nl/~lindeboo

NEW ZEALAND
EBSCO NZ Ltd.
Private Mail Bag 99914
New Market
Auckland
Tel: (64 9) 524-8119
Fax: (64 9) 524-8067

NIGERIA
University Press Limited
Three Crowns Building Jericho
Private Mail Bag 5095
Ibadan
Tel: (234 22) 41-1356
Fax: (234 22) 41-2056

NORWAY
NIC Info A/S
Book Department, Postboks 6512 Etterstad
N-0606 Oslo
Tel: (47 22) 97-4500
Fax: (47 22) 97-4545

PAKISTAN
Mirza Book Agency
65, Shahrah-e-Quaid-e-Azam
Lahore 54000
Tel: (92 42) 735 3601
Fax: (92 42) 576 3714

Oxford University Press
5 Bangalore Town
Sharae Faisal
P.O. Box 13033
Karachi-75350
Tel: (92 21) 446307
Fax: (92 21) 4547640
E-mail: ouppak@TheOffice.net

Pak Book Corporation
Aziz Chambers 21, Queen's Road
Lahore
Tel: (92 42) 636 3222; 636 0885
Fax: (92 42) 636 2328
E-mail: pbc@brain.net.pk

PERU
Editorial Desarrollo SA
Apartado 3824, Lima 1
Tel: (51 14) 285380
Fax: (51 14) 286628

PHILIPPINES
International Booksource Center Inc.
1127-A Antipolo St, Barangay, Venezuela
Makati City
Tel: (63 2) 896 6501; 6505; 6507
Fax: (63 2) 896 1741

POLAND
International Publishing Service
Ul. Piekna 31/37
00-677 Warzawa
Tel: (48 2) 628-6089
Fax: (48 2) 621-7255
E-mail: books%ips@ikp.atm.com.pl
URL: http://www.ipscg.waw.pl/ips/export/

PORTUGAL
Livraria Portugal
Apartado 2681, Rua Do Carmo 70-74
1200 Lisbon
Tel: (1) 347-4982
Fax: (1) 347-0264

ROMANIA
Compani De Librarii Bucuresti S.A.
Str. Lipscani no. 26, sector 3
Bucharest
Tel: (40 1) 613 9645
Fax: (40 1) 312 4000

RUSSIAN FEDERATION
Isdatelstvo <Ves Mir>
9a, Kolpachniy Pereulok
Moscow 101831
Tel: (7 095) 917 87 49
Fax: (7 095) 917 92 59

SINGAPORE, TAIWAN, MYANMAR, BRUNEI
Ashgate Publishing Asia Pacific Pte. Ltd.
41 Kallang Pudding Road #04-03
Golden Wheel Building
Singapore 349316
Tel: (65) 741-5166
Fax: (65) 742-9356
E-mail: ashgate@asianconnect.com

SLOVENIA
Gospodarski Vestnik Publishing Group
Dunajska cesta 5
1000 Ljubljana
Tel: (386 61) 133 83 47; 132 12 30
Fax: (386 61) 133 80 30
E-mail: repansekj@gvestnik.si

SOUTH AFRICA, BOTSWANA
For single titles:
Oxford University Press Southern Africa
Vasco Boulevard, Goodwood
P.O. Box 12119, N1 City 7463
Cape Town
Tel: (27 21) 595 4400
Fax: (27 21) 595 4430
E-mail: oxford@oup.co.za

For subscription orders:
International Subscription Service
P.O. Box 41095
Craighall
Johannesburg 2024
Tel: (27 11) 880-1448
Fax: (27 11) 880-6248
E-mail: iss@is.co.za

SPAIN
Mundi-Prensa Libros, S.A.
Castello 37
28001 Madrid
Tel: (34 1) 431-3399
Fax: (34 1) 575-3998
E-mail: libreria@mundiprensa.es
URL: http://www.mundiprensa.es/

Mundi-Prensa Barcelona
Consell de Cent, 391
08009 Barcelona
Tel: (34 3) 488-3492
Fax: (34 3) 487-7659
E-mail: barcelona@mundiprensa.es

SRI LANKA, THE MALDIVES
Lake House Bookshop
100, Sir Chittampalam Gardiner Mawatha
Colombo 2
Tel: (94 1) 32105
Fax: (94 1) 432104
E-mail: LHL@sri.lanka.net

SWEDEN
Wennergren-Williams AB
P.O. Box 1305
S-171 25 Solna
Tel: (46 8) 705-97-50
Fax: (46 8) 27-00-71
E-mail: mail@wwi.se

SWITZERLAND
Librairie Payot Service Institutionnel
Côtes-de-Montbenon 30
1002 Lausanne
Tel: (41 21) 341-3229
Fax: (41 21) 341-3235

ADECO Van Diemen EditionsTechniques
Ch. de Lacuez 41
CH1807 Blonay
Tel: (41 21) 943 2673
Fax: (41 21) 943 3605

THAILAND
Central Books Distribution
306 Silom Road
Bangkok
Tel: (66 2) 235-5400
Fax: (66 2) 237-8321

TRINIDAD & TOBAGO AND THE CARRIBBEAN
Systematics Studies Ltd.
St. Augustine Shopping Center
Eastern Main Road, St. Augustine
Trinidad & Tobago, West Indies
Tel: (868) 645-8466
Fax: (868) 645-8467
E-mail: tobe@trinidad.net

UGANDA
Gustro Ltd.
P.O. Box 9997, Madhvani Building
Plot 16/4 Jinja Rd.
Kampala
Tel: (256 41) 251 467
Fax: (256 41) 251 468
E-mail: gus@swiftuganda.com

UNITED KINGDOM
Microinfo Ltd.
P.O. Box 3, Alton, Hampshire GU34 2PG
England
Tel: (44 1420) 86848
Fax: (44 1420) 89889
E-mail: wbank@ukminfo.demon.co.uk
URL: http://www.microinfo.co.uk

VENEZUELA
Tecni-Ciencia Libros, S.A.
Centro Cuidad Comercial Tamanco
Nivel C2, Caracas
Tel: (58 2) 959 5547; 5035; 0016
Fax: (58 2) 959 5636

ZAMBIA
University Bookshop, University of Zambia
Great East Road Campus
P.O. Box 32379
Lusaka
Tel: (260 1) 252 576
Fax: (260 1) 253 952